BIOCHEMISTRY OF PARASITIC PROTOZOA

BIOCHEMISTRY OF PARASITIC PROTOZOA

W. E. GUTTERIDGE

*Senior Lecturer in Biology, University of Kent
at Canterbury, England*

G. H. COOMBS

*Lecturer in Zoology, University of Glasgow
Scotland*

UNIVERSITY PARK PRESS
Baltimore • London • Tokyo

First published 1977 by
The Macmillan Press Ltd
London and Basingstoke

Published in North America by
UNIVERSITY PARK PRESS
Chamber of Commerce Building
Baltimore, Maryland 21202

Printed in Great Britain

Library of Congress Cataloging in Publication Data

Gutteridge, W. E.
   Biochemistry of Parasitic Protozoa.

   Bibliography: p.
   Includes index.
   1.   Protozoa, Pathogenic—Physiology.   I.   Coombs,
G. H., joint author.   II.   Title. |DNLM: 1.   Protozoa—
Pathogenicity.   2.   Antiprotozoal agents.   QZ85 G985b|
QR251.G87   1977   616.01′6   77–6355
ISBN 0–8391–0886–9

Contents

Preface

In recent years there have been several introductory texts dealing with different aspects of parasitic protozoa, but none has considered the subject biochemically. This book is an attempt to remedy the situation.

It is a résumé of our present (literature surveyed to December 1976) knowledge of the biochemistry of parasitic protozoa and the mode of action of the drugs useful in the treatment of the diseases they cause. The information is presented in a comparative manner; in each section, the protozoan metabolism is contrasted with that of the mammalian host and the differences between the two, which have been or might in future be exploited for chemotherapeutic purposes, are emphasised. Where the present situation is unclear because of the existence of conflicting information, only the conclusions from a critical appraisal are presented to avoid confusing the reader with unnecessary detail. The text makes it clear where this has been done. In addition to stating the known facts, an attempt is made also to point out the present deficiencies in our knowledge. To keep the length of the book within bounds and to reflect current research interests, only genera with species parasitic in man or domestic animals have been considered. To avoid confusing readers new to the field, no attempt has been made to credit the work described to individual scientists. We hope our colleagues will not be offended by this decision. Suggestions for further reading are included at the end of each chapter, and, where possible, these have been restricted to readily available books and review articles. Advice on how to follow the biochemical protozoology literature is given in appendix D.

The book is written in such a way that little prior knowledge of either biochemistry or protozoology is necessary and it is aimed at the final-year undergraduate, first-year postgraduate level. Thus it should be useful not only to intending protozoologists but also to students in the related fields of biochemistry, biology, medicine, microbiology, parasitology, veterinary medicine and zoology.

We are indebted to the following persons for the provision of photographs: Dr J. Alexander for figure 1.3c, Mr B. Cover for figures 1.3a, 1.3b, 1.3e, 1.3g and 1.3i; Dr V. S. Latter for figure 1.3f; Dr N. McHardy for figure 1.3j; Mr R. Newsam for figure 1.1; Professor G. Riou for figure 6.10; and Professor K. Vickerman for figure 7.5. We thank our colleagues in Canterbury and Glasgow for helpful discussions, Dr P. I. Trigg and Dr J. Williamson for reading parts of the manuscript and Mrs Sue Cover for her excellent typing of it.

W. E. Gutteridge
G. H. Coombs

1 Introduction to Parasitic Protozoa

1.1 What are Protozoa?

Protozoa may be regarded either as a phylum within the animal kingdom or as a group of microorganisms within the Protista having the basic characteristics of animal cells. In addition, they may be regarded either as unicellular or noncellular organisms. There has been considerable controversy as to which of each pair of definitions is more correct. Such arguments, however, are only of passing interest to the biochemist and therefore will not be pursued.

The basic cellular organisation of protozoa is of the eukaryotic type, that is the cell contents are delineated into a large number of membrane-bound organelles such as nuclei, mitochondria, Golgi apparatus, lysosomes and food vacuoles (figure 1.1). This organisation is quite distinct from that found in prokaryotic microorganisms, for instance bacteria, which lack membrane-bound organelles but is similar to that found both in other lower eukaryotes such as algae and fungi and also in the cells of higher plants and animals. Absence of a mechanically rigid cell wall external to the plasma membrane distinguishes protozoan cells from those of algae, fungi and higher plants and underlies their similarity to those of multicellular animals.

1.2 Classification of the Phylum Protozoa

The primary classification of protozoa into four subphyla is made principally on the type of locomotory organelle (flagellum, cilium, pseudopodium or none of these) (figure 1.2). The subphylum Sarcomastigophora is further subdivided in a similar way. Most of the secondary classification, however, is based on variation of morphological features in either vegetative or sexual stages.

1.3 Important Genera of Parasitic Protozoa

There are twelve genera of parasitic protozoa which are important to man because they cause disease either in him or his domestic animals (see figure 1.2). Note that all twelve fall within the subphyla Sarcomastigophora and Sporozoa; none occurs in the subphyla Ciliophora and Cnidospora although many species in these two groups are parasitic. Of the twelve, only

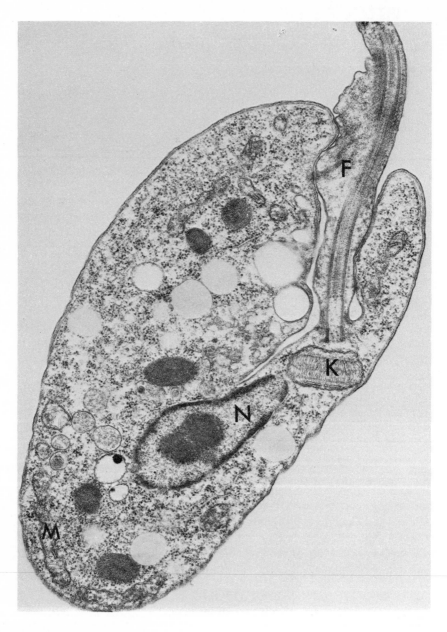

Figure 1.1 Electron micrograph of *Crithidia fasciculata* (×15 000). N, nucleus; K, kinestplast; F, flagellum; M, mitochondrion.

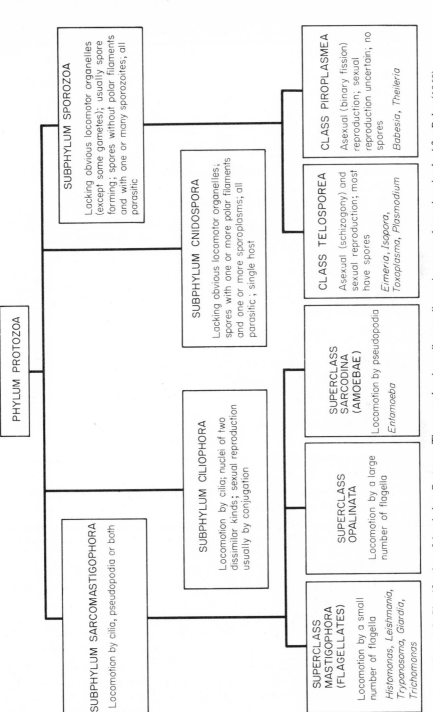

Figure 1.2 Classification of the phylum Protozoa. The examples given all cause disease in man or domestic animals. After Baker (1969).

Trypanosoma, Leishmania, Trichomonas (flagellates), *Entamoeba (amoeba),*
Eimeria, Toxoplasma and *Plasmodium* (sporozoans) have been
studied biochemically so far although work is now beginning on two other
sporozoans *Babesia* and *Theileria*. This book, therefore, is concerned solely
with these nine genera. Information about them is summarised in the photomicrographs in figure 1.3 and in tables 1.1 and 1.2.

1.4 Genus *Trypanosoma*

This genus, together with the genus *Leishmania*, forms part of the order
Kinetoplastida. Members of this order are characterised by the presence at the
base of the flagellum of a DNA-containing organelle—the kinetoplast.
Different morphological forms of these flagellates have been recognised on the
basis of the position of the kinetoplast relative to the nucleus (figure 1.4). A
proliferation of the membranes associated with the kinetoplast into tubular-like
structures forms the single mitochondrion of these organisms. Structural rigidi-
ty in the organism is provided by subpellicular microtubules which lie just
below the plasma membrane. There is usually only one flagellum which
emerges from the flagellar pocket and may be attached to the body wall to
form an undulating membrane.

For reasons related both to the diseases they cause and to their morphology
and biochemistry, trypanosomes are classified into salivarian (mainly African)
and stercorarian species.

The salivarian species include *T. rhodesiense* and *T. gambiense,* which
cause sleeping sickness in man, and *T. brucei, T. congolense* and *T. vivax*
which cause nagana and similar diseases in cattle. These diseases are restricted
to the African continent, although there are related diseases, caused by other
species, which occur elsewhere. Constant monitoring and rapid treatment of
people living in areas in which trypanosomiasis is endemic have prevented
serious outbreaks of the disease in recent years. It is estimated that probably
there are no more than 250 000 cases in Africa at the present time. The disease
in cattle, however, is still of great economic importance since it causes the
deaths of 3 000 000 head per year and effectively prevents the use of vast areas
for ranching.

In the mammalian host, salivarian trypanosomes live in the body fluids including the blood and in late stages the cerebrospinal fluid, where they grow
and divide by binary fission. Transmission in most species involves insect vectors, usually tsetse flies (*Glossina* spp.). The complex life cycle of *T. brucei,* a
species similar to *T. gambiense* and *T. rhodesiense* but not pathogenic to man
and therefore much used in research, is illustrated diagrammatically in figure 1.5.
There are two distinct blood trypomastigote stages in the mammal (long
slender and short stumpy forms which are, as their names imply, long and thin
and short and fat, respectively) and many developmental forms of the parasite
in the insect vector, features that are referred to later in the book. The vector

Table 1.1 Some genera of parasitic protozoa that cause disease
in man and domestic animals

Genus	Type	Diseases	Tissue/ organ infected	Transmission	Geography
Trypanosoma (African species)	Flagellate	Trypanosomiasis (sleeping sickness) in man; nagana in cattle; others	Blood/ lymph	Tsetse flies	Tropical Africa
Trypanosoma (*T. cruzi*)	Flagellate	Chagas' disease in man	Blood/ muscle of heart and gut	Reduviid bugs	Tropical South America
Leishmania	Flagellate	Leishmaniasis in man: cutaneous (oriental sore) and visceral (kala-azar)	Macrophages of skin or viscera	Sand flies	Tropics
Trichomonas	Flagellate	Trichomoniasis (trichomonal vaginitis) in man	Urinogenital system	Sexual	World wide
Entamoeba	Amoeba	Amoebiasis (amoebic dysentery)	Gut lumen/ viscera (especially liver)	Cysts	Mainly tropics
Eimeria	Sporozoan	Coccidiosis in chickens	Gut epithelial cells	Cysts	World wide
Toxoplasma	Sporozoan	Toxoplasmosis in man	Macrophages	Cysts from infected cats	World wide
Plasmodium	Sporozoan	Malaria in man	Liver, red blood cells	Mosquitoes	Tropics
Babesia	Sporozoan	Babesiasis in cattle	Red blood cells	Ticks	World wide
Theileria	Sporozoan	Theileriasis (for example east coast fever) in cattle	Red blood cells, lymphocytes	Ticks	World wide (east coast fever in Africa)

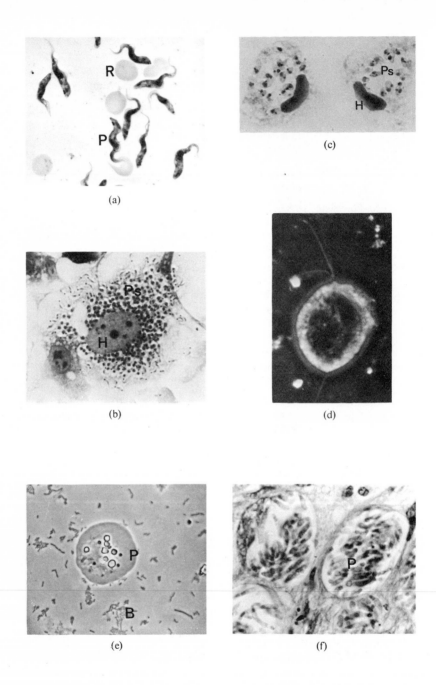

(a)

(c)

(b)

(d)

(e)

(f)

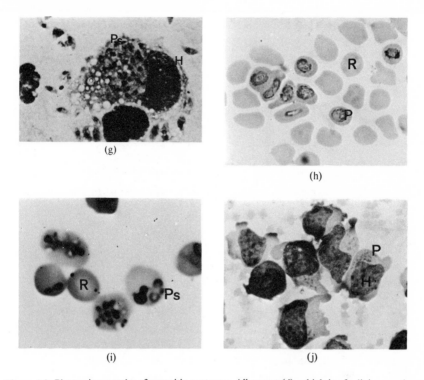

Figure 1.3 Photomicrographs of parasitic protozoa. All except (d), which is of a living specimen viewed by phase-contrast microscopy, are of fixed and stained preparations viewed by light microscopy.

(a) Trypomastigote forms of *Trypanosoma congolense* in smear of mouse blood (×4000). (b) Intracellular stages of *Trypanosoma cruzi* in a human heart tissue cell (×800). Note that trypomastigote forms are beginning to differentiate from amastigote forms near the perifery of the cell. (c) Amastigote forms of *Leishmania mexicana* in a peritoneal macrophage (×1000). (d) *Trichomonas vaginalis* from culture (×1500). (e) Trophozoite of *Entamoeba histolytica* from monaxienic culture (×1500). (f) Schizonts of *Eimeria tenella* from the caecum of a chicken (×1000). (g) Trophozoites of *Toxoplasma gondii* in a macrophage of a peritoneal smear (×1000). (h) Trophozoites of *Plasmodium knowlesi* in Rhesus monkey red blood cells (×1000). (i) Intraerythrocytic stages of *Babesia rodhaini* in mouse red blood cells (×1500). (j) Macroschizonts of *Theileria parva* in lymphocytes from the lymph node of a cow (×800).

P, parasite; Ps, parasites; R, red blood cells; H, host cell nucleus; B, bacteria.

Table 1.2 Parasitic protozoa causing disease, and useful laboratory species

Genus	Species	Hosts
Trypanosoma (salivaria)	*rhodesiense*	Man, game animals, cattle
	gambiense	Man, game animals, cattle
	brucei	Game animals, cattle
	evansi	Camels
	equinum	Horses
	equiperdum	Horses
	congolense	Game animals, cattle
	vivax	Game animals, cattle
Trypanosoma (stercoraria)	*cruzi*	Man, wild animals
	lewisi	Rat
Leishmania	*donovani*	Man (viscera), dogs, rodents, etc.
	tropica	Man (skin), dogs, rodents, etc.
	braziliensis	Man (skin), rodents, etc.
	mexicana	Man (skin), rodents, etc.
Trichomonas	*vaginalis*	Man
	foetus	Cattle
	gallinae	Birds
Entamoeba	*histolytica*	Man, primates, dogs, cats, etc.
	invadens	Reptiles
Eimeria	*tenella*	Chickens
	necatrix	Chickens
Toxoplasma	*gondii*	Man, cats, dogs, cattle, etc.
Plasmodium	*vivax*	Man
	malariae	Man
	ovale	Man
	falciparum	Man, *Aotus* monkeys
	knowlesi	Rhesus monkeys
	berghei	Rodents
	vinckei	Rodents
	chabaudi	Rodents
	gallinaceum	Chickens
	lophurae	Ducks
Babesia	*bovis*	Cattle
	rodhaini	Rodents
	microti	Rodents
Theileria	*parva*	Cattle

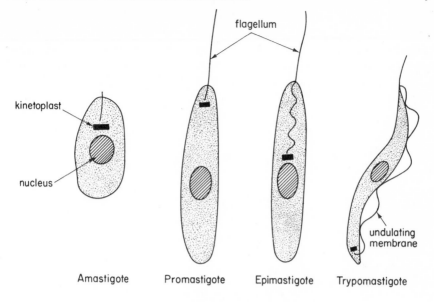

Figure 1.4 Morphological forms of kinetoplastid flagellates.

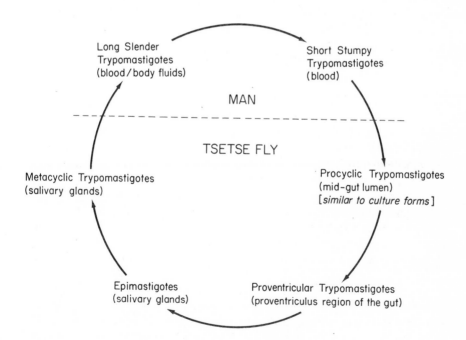

Figure 1.5 The life cycle of *Trypanosoma brucei*.

becomes infected while taking a blood meal and once the salivary glands have become infected is able to pass on the infection during subsequent feeding.

Crithidia fasciculata is a parasite of insects which can be cultured easily *in vitro* and often is used as a model of salivarian trypanosomes. Morphological-ly, it is distinct from all developmental stages of *T. brucei* but is most similar to the epimastigote stage.

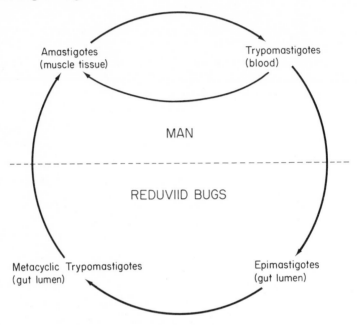

Figure 1.6 The life cycle of *Trypanosoma cruzi*.

The most important stercorarian species is *T. cruzi* which causes Chagas' disease in man. There is no equivalent disease in domestic animals. Although the parasite and the vectors that transmit it occur on the American continent 40° north and south of the equator, the disease appears to be restricted to the South American subcontinent. Twelve million people are known to be infected and at least 35 million live in regions where the disease is endemic. These are mostly the poorer rural areas where the mud-plastered walls of the houses, once they have become dried and cracked, provide optimal environments for the breeding of the reduviid bugs which transmit the infection.

The life cycle is illustrated diagrammatically in figure 1.6. The parasite grows and divides in the amastigote form in the muscle tissue of the heart and alimentary tract of man. The blood trypomastigote stages do not divide but seem only to spread the infection around the body and to infect the insect vec-tor during its feeding. The parasite develops in the gut of the insect mainly in the epimastigote form. There is no infection of the salivary glands. It is transmitted back to man because the bugs defecate during feeding. The infective metacyclic trypomastigotes from the faeces enter the new host through skin lesions.

1.5 Genus *Leishmania*

There are 3 species in this genus of kinetoplastid flagellates causing important diseases in man. *L. donovani,* which causes visceral leishmaniasis or kala-azar, occurs throughout the warmer parts of Asia, the Mediterranean coasts, North and East Africa and South America. This parasite, as with all other *Leishmania* species, grows intracellularly in the macrophages of the body in the amastigote form. Infected cells occur in all tissues, including the blood. The disease takes a long time to run its course but is often fatal. *L. tropica* causes

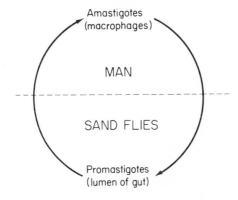

Figure 1.7   The life cycle of *Leishmania donovani.*

cutaneous leishmaniasis or oriental sore. The disease has a more limited distribution geographically than kala-azar and the parasites themselves are restricted to ulcers in the skin. It usually cures itself. *L. braziliensis,* which occurs in Brazil, is sometimes regarded as a subspecies of *L. tropica.* However, a disease it may cause, mucocutaneous leishmaniasis, is much more severe.

As with the trypanosomes, the life cycle of *Leishmania* is complex and involves an insect vector, sand flies (figure 1.7). Growth and division occurs in the amastigote form in the mammal and in the promastigote form in the gut of the vector. Infection is passed on during feeding because the sand flies (*Phlebotomus* spp.) tend to regurgitate. There is no infection of the salivary glands.

1.6 Genus *Trichomonas*

These flagellates are ovoid with three to five free anterior flagella and one recurrent flagellum attached to the body to form an undulating membrane.

There is a prominent skeletal axostyle comprising longitudinally arranged microtubules. Mitochondria are absent but unique organelles called hydrogenosomes occur. Only one developmental stage is known.

Two species are of importance to man. *T. vaginalis* occurs mainly in the female vagina and male urethra and prostate gland and is common throughout the world. Infection rates as high as 30 per cent have been recorded in some areas. The parasite often is present without causing any symptoms in men, but an extremely unpleasant but non-fatal condition—trichomonal vaginitis—can occur in women. Reproduction is solely by binary fission (no sexual process is known) and the parasite is transmitted sexually. *T. foetus* is a parasite of cattle. It lives in the vagina and uterus of the female and can cause abortion. It is transmitted sexually and occurs throughout the world, although, in recent years, increasing use of artificial insemination techniques has reduced its occurrence.

1.7 Genus *Entamoeba*

These typical amoeboid protozoa can be recognised readily by the peripheral arrangement of the nucleoprotein in their nuclei and the absence of mitochondria. The only species of importance to man is *E. histolytica* which can cause amoebic dysentery, a disease that is confined mainly to the tropics. *E. histolytica* usually lives in the lumen of the gut as a harmless commensal but occasionally, for reasons that are not fully understood, the trophozoites penetrate the wall of the gut, form ulcers and cause severe tissue damage and blood loss. Often they also spread through the blood to other body organs such as the liver where they cause amoebic abscesses which, if not treated, can be fatal. The disease is spread by four-nucleate cysts which are voided in the faeces and can easily contaminate food if standards of hygiene are not high. The cysts pass through the stomach and set up infection in the intestine.

1.8 Genus *Eimeria*

This sporozoan genus contains a large number of species parasitising vertebrate animals of all classes throughout the world, although none infects man. They are often referred to, along with members of the genus *Isospora*, as the 'coccidia'. A typical species is *Eimeria tenella* which infects the chicken. It grows intracellularly in the epithelial cells of the caecum as a trophozoite stage, later undergoing schizogony (a form of multiple mitosis) to form a large number of merozoites. These are released as the host cell lyses and infect fresh cells. The wall of the caecum is badly damaged as a result of this development, leading to severe blood and fluid loss which can result in death.

The life cycle is complex and involves sexual reproduction but no insect vector. Instead, resistant oocysts are produced which are voided in the faeces and can easily spread the infection by contamination of food (figure 1.8).

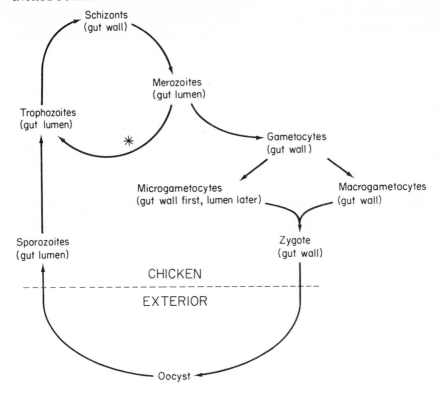

Figure 1.8 The life cycle of *Eimeria*. *Not in all species.

1.9 Genus *Toxoplasma*

The only important species is *T. gondii* which can cause toxoplasmosis in man. The parasite is widespread throughout the world (about 30 per cent of the population are infected) but in most people the infection is latent and asymptomatic so that the number of clinical cases in a year is small (approximately 200 in the United Kingdom). In addition, however, the parasite can affect the development of the foetus and cause abortion. The trophozoite stage of the organism has, like the *Eimeria* trophozoite, an unremarkable structure. It can develop in the macrophages of almost all the tissues of the body, but it shows a particular preference for the central nervous system, the musculature and the lungs.

Until recently, details of the life cycle were not known. However, in 1969, it was shown that a developmental cycle occurs in the domestic cat, though probably not in man or any other animal. Gametes and oocysts are produced as in *Eimeria* and the oocysts are voided in the faeces. Other cats, man and domestic animals are infected by eating food contaminated with these oocysts. These observations provide an explanation for both the transmission of toxoplasmosis in man and also the high incidence of *Toxoplasma* in domestic animals and especially in the cat.

1.10 Genus *Plasmodium*

There are four species of the genus *Plasmodium* which cause malaria in man. *P. vivax*, *P. malariae* and *P. ovale* cause a relatively mild disease, which, although debilitating, is rarely fatal, but can persist for a number of years. In contrast, *P. falciparum* causes an acute disease in man which, if untreated, is often fatal. In spite of intensive efforts at eradication over the past thirty years, it has been estimated that at least 480 million people still live in areas of the world where malaria is endemic. These areas include most of tropical Africa (where malaria is a major cause in over 1 million deaths annually) and South

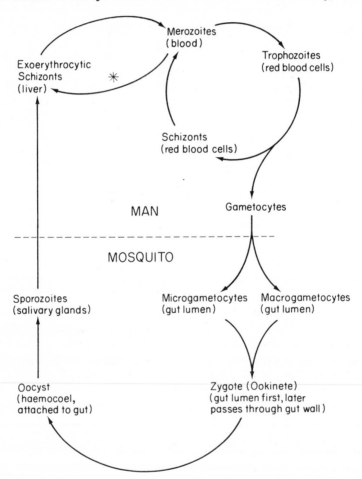

Figure 1.9 The life cycle of *Plasmodium*. *Not in all species of malaria.

America, large parts of the Middle East, parts of India and Sri Lanka and the whole of South-east Asia.

In the vertebrate host the malaria parasite lives mainly in the erythrocytes where it digests the haemoglobin as a source of amino acids with the concomitant production of pigment. The morphology varies with both the stage of development and with the species but in all cases, the intraerythrocytic parasite, which is in a vacuole surrounded by host-cell membrane, is a small trophozoite that grows and divides by schizogony (multiple fission). The disease is transmitted by female *Anopheles* mosquitoes and the life cycle is basically similar to that of *Eimeria* (figure 1.9). Fusion of the gametocytes occurs in the mosquito gut and the resultant oocyst produces large numbers of sporozoites which infect the salivary glands. Transfer of these to man occurs during feeding. They form exoerythrocytic schizonts in the liver which later initiate the typical infection in the blood cells. The development of the intraerythrocytic schizonts and their rupture to release merozoites tends to be synchronous, which accounts for the periodic fevers so characteristic of malaria. The time elapsing between successive fevers is therefore the same as the duration of the intraerythrocytic cycle—72 h for *P. malariae* and 48 h for the other three species.

1.11 Genus *Babesia*

Members of this genus cause a number of diseases (collectively known as babesiasis) of veterinary importance. They are responsible for the deaths of 250 000 cattle each year. The genus has a world-wide distribution and also includes many species of low pathogenicity which are parasites of wild animals.

Babesia live in the erythrocytes of the host but, unlike malaria parasites from which they can be distinguished by the absence of pigment, they are not surrounded by host-cell membranes. Organisms are rounded or oval and division stages are often seen which give rise to two or four merozoites. Transmission involves the tick as vector. The life cycle within the tick is complicated and not understood fully, but eventually infective forms are produced in the salivary glands and are transmitted during feeding (figure 1.10).

1.12 Genus *Theileria*

This genus is closely related to *Babesia*. It includes a number of species that cause diseases of veterinary importance. The best known is east coast fever which is responsible for the deaths of 500 000 cattle each year in Africa and is caused by *T. parva*. In the mammalian host, the parasites develop initially in lymphocytes as macroschizonts, causing host lymphocyte proliferation. Parasites divide in synchrony and thus the number of infected cells increases enormously. Later in the infection, microschizonts are formed. These give rise to micromerozoites which invade the red blood cells. The erythrocyte stages are infectious to ticks which act as vectors of the disease. Transmission to a new mammalian host is by parasites in the salivary glands (figure 1.11).

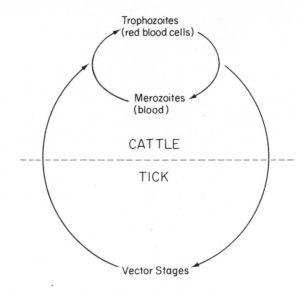

Figure 1.10 The life cycle of *Babesia*.

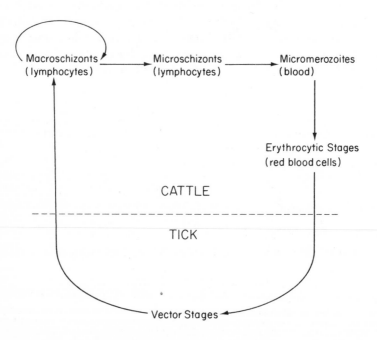

Figure 1.11 The life cycle of *Theileria*.

1.13 Further Reading

Adam, K. M. G., Paul, J. and Zaman, V. (1971). *Medical and Veterinary Protozoology.*
  Churchill Livingstone, Edinburgh and London
Baker, J. R. (1969). *Parasitic Protozoa.* Hutchinson University Library, London
Sleigh, M. (1973). *The Biology of Protozoa.* Edward Arnold, London
Vickerman, K. and Cox, F. E. G. (1967). *The Protozoa.* John Murray, London

2 Control of Diseases caused by Parasitic Protozoa

2.1 Introduction

Prevention (prophylaxis) and treatment of diseases caused by microorganisms has been attempted in four ways:

(1) improvements in housing, sanitation and hygiene
(2) control of vectors
(3) development of vaccines
(4) development of drugs

The first three methods are useful only for prophylaxis; drugs are useful both for prophylaxis and for treating existing infections. The application of these techniques, often in combination, has led to the control of most of the viral, bacterial and fungal diseases of man and domestic animals.

Parallel attempts to control protozoan diseases have also been carried out. A biochemical text, however, is not the place for a detailed discussion of the value of housing improvements and vector control so that these topics will not be discussed further. This is not to say that their use should be discounted. For example, vector control, more than any other, has led to the eradication of malaria from many parts of the world where it was previously endemic. Similarly, it is generally accepted that the problem of Chagas' disease in South America could be solved, if finance was available, simply by building brick instead of mud-walled houses and so eliminating the habitat of the insect vector. This chapter, however, will concentrate on surveying the present state of development of techniques of vaccination and chemotherapy for the control of protozoan diseases.

2.2 Vaccination

Vaccination (or immunisation) as a means of controlling infectious diseases dates back to the middle of the fifteenth century when the Chinese used dried crusts from smallpox lesions to induce a, hopefully, short mild infection which would confer resistance to a subsequent severe infection. The use of cowpox crusts in 1791 was a major advance making the procedure safer and led eventually in the second half of the nineteenth century first to the use of naturally-occurring attenuated strains and later the development of methods of artificial attenuation (for example, formalin treatment and X-irradiation). The object of attenuation is to destroy the infectivity of the microorganism and thus its ability to cause disease, but to leave its antigenicity unimpaired. The treatment, usually by intramuscular or subcutaneous injection, of a patient with an attenuated strain of the infective organism will render him immune to subsequent infection without inducing disease at the vaccination stage. However, attenuation sometimes reduces antigenicity. In the experimental situation, this problem is overcome by the use of adjuvants (for example, Freund's) which boost the immune response, but most of these are too toxic for use in man.

The success of vaccination procedures is primarily due to the phenomenon of immunological memory. The immune response to a previously encountered antigen (secondary response) occurs much more quickly and to a higher level than it does to a new antigen (primary response) due to the presence of memory lymphocytes which can respond to that antigen. Thus the infection is cleared rapidly, usually before clinical symptoms have occurred.

Such procedures have been valuable in the control of many bacterial and viral diseases. However, although a vaccine for the veterinary disease babesiasis is in use in Australia and South-east Asia, and much experimental work is in progress so that the situation will possibly alter radically in the next few years, there is at present no other commercially available vaccine for any protozoan disease of man or his domestic animals. The reasons for this might be considered to be outside the scope of this book since essentially they are problems of immunology and culture. They will be discussed briefly, however, since the underlying problems are in some cases physiological and biochemical.

The mammalian blood stream offers parasites a rich supply of nutrients and oxygen and facilities for the rapid removal of waste products. The presence of the parasites there, however, inevitably provokes the immune response of the host. Biochemists working on parasitic protozoa require large amounts of material and so they usually work with acute infections where large numbers of parasites, which often rapidly overwhelm the host, are present. Natural infections, however, are often chronic and involve the survival of small numbers of parasites over long periods of time. A number of factors seem to underlie the ability of the parasite in these conditions to escape the immune response of the host, including antigenic variation (*Trypanosoma*, *Plasmodium*, *Babesia*), production of soluble blocking antigens (possibly *Trypanosoma*, *Plasmodium*), coating of the protozoan by host antigen (*T. vivax*, *T. lewisi*), anatomical seclusion (*T. cruzi*, *Leishmania*, *Trichomonas*, *Entamoeba*, *Eimeria*, *Toxoplasma*, *Plasmodium*), immunosuppression (*Trypanosoma*, *Plasmodium*), and acquired tolerance (*Leishmania*). These factors can all contribute to the problems of developing protozoan vaccines. In addition, our inability to culture most parasitic protozoa *in vitro* (section 3.2 and appendix B) makes mass production of any successful vaccine seem at present a major undertaking.

2.3 Chemotherapy

2.3.1 Introduction

Chemotherapy as a means of controlling infectious diseases was developed by
Ehrlich and his colleagues in Germany towards the end of the last century.
Ehrlich at that time was studying the staining properties of a number of dyes
for protozoa such as trypanosomes and malaria parasites. He conceived the
idea that it ought to be possible to find dyes (drugs) which would selectively
destroy pathogens but leave host cells unimpaired. His approach was
successful with the protozoa. By 1930, a number of drugs were known which
would help in the control of several diseases including sleeping sickness,
malaria and amoebiasis (organic arsenicals and suramin, plasmoquine and
stovarsol, respectively). It is not always appreciated that the approach was far
less successful with antibacterial drugs. It needed the discovery of anti-
metabolites (for example, sulphonamides) and antibiotics (for example,
penicillins) to open up this area of chemotherapy. However, the original approach adopted by Ehrlich of achieving drug selectivity through differential
toxicity towards host and pathogen is still basic to all modern drug development. Thus, it can truly be said that the story of the development of antiprotozoan drugs is the story of the development of chemotherapy itself.

Today, in contrast to the situation with vaccines, all the protozoan diseases
under consideration (section 1.3) can be controlled, at least in part, by drugs,
with the exception of east coast fever (but see subsection 2.3.10). For many of
them, chemoprophylactic agents are also available. The drugs most frequently
used are listed in table 2.1. Their chemical structures and biochemical modes of
action are considered in chapter 9.

2.3.2 Trypanosomiasis (except Chagas' disease)

The early stages of sleeping sickness in man are controlled by suramin and occasionally by pentamidine and berenil although the former is used mainly for
chemoprophylaxis and the latter has not been cleared by the manufacturers for
use in man. None of these three drugs will cross the cerebrospinal membrane
and so control of the late stages of the disease, in which trypanosomes are present in the cerebrospinal fluid (see section 1.4), rests with arsenicals such as
tryparsamide and Mel B. The chemotherapeutic index (ratio of toxic to
curative dose) of the latter compound is low and so it is given only to hospitalised
patients.

Nagana in cattle and related diseases in horses and camels have in the past
been treated by drugs such as ethidium, dimidium and antrycide but problems
of resistance mean that, at present, the drug of choice is berenil.

2.3.3 Chagas' disease

There is as yet no effective chemotherapy for all stages of this disease. Lampit, however, has just been introduced into South America and seems effective in controlling the early acute phase of the disease caused by an abundance of trypomastigotes in the blood. There is doubt, however, about its ability to cure the later, chronic stage, where most of the parasites are intracellular in the muscle tissue of the heart and alimentary tract. The doses and periods of treatment required are more than most patients can tolerate. Radanil is at present undergoing clinical trials, but it is too early to judge whether they will be successful.

2.3.4 Leishmaniasis

Pentostam and occasionally amphotericin B and pentamidine are used for the treatment of the cutaneous form of the disease and pentamidine for the visceral form. Lampit is undergoing clinical trials.

2.3.5 Trichomoniasis and amoebiasis

Metronidazole and other 5-nitroimidazoles are now the drugs of choice for both diseases. Since trichomoniasis is sexually transmitted, it is customary to treat both partners.

2.3.6 Coccidiosis

Chickens kept in battery conditions (as most now are) are normally given continuous chemoprophylactic treatment by mixing a drug or combination of drugs with the diet. Obviously, in these conditions some chickens will obtain subcurative doses and this can lead to the rapid development of drug-resistant strains of parasite. Thus the average life of an anticoccidial drug in continuous usage is about two years. To avoid this problem, changes in the drug used are made frequently. This is the reason why there are so many drugs listed in table 2.1.

2.3.7 Toxoplasmosis

Pyrimethamine on its own or in combination with dapsone are the treatments of choice.

2.3.8 Malaria

Chloroquine and pyrimethamine on its own or in combination with dapsone are used for chemoprophylaxis and chloroquine or quinine (especially for chloroquine-resistant strains) for treatment of blood infections. Chloroquine has no effect on liver stages and so, in infections caused by *Plasmodium vivax*, *P. malariae* and *P. ovale*, use of primaquine or another 8-aminoquinoline is recommended to prevent relapse. For other cases of drug resistance,

Table 2.1 Drugs in current use for the chemotherapy and chemoprophylaxis of diseases caused by parasitic protozoa

Disease	Drug	Chemistry	Formula* given in figure:	Clinical use
Trypanosomiasis	Tryparsamide	Arsenical	9.1	Late treatment in man (*T. gambiense* only)
	Mel B	Arsenical	9.1	Late treatment in man
	Suramin	Sulphated naphthylamine	9.3	Early treatment in man
	Pentamidine	Diamidine	9.18	Prophylaxis in man
	Ethidium	Phenanthridine	9.15	Treatment in cattle etc.
	Dimidium	Phenanthridine	9.15	Treatment in cattle etc.
	Antrycide	Aminoquinaldine	9.19	Treatment in cattle etc.
	Berenil	Diamidine	9.18	Treatment in cattle etc.
Chagas' Disease	Lampit	5-Nitrofuran	9.17	Treatment of acute stage in man
	Radanil	2-Nitroimidazole	9.17	Clinical trials of acute and chronic stages in man
Leishmaniasis	Pentostam	Antimonial	9.2	Treatment of cutaneous disease
	Amphotericin B	Polyene	9.8	Treatment of cutaneous disease
	Pentamidine	Diamidine	9.18	Treatment of visceral disease
Trichomoniasis and Amoebiasis	Metronidazole	5-Nitroimidazole	9.17	Treatment in man
Coccidiosis	Amprolium/ethopabate	Thiamine analogue/2-substituted *p*-aminobenzoic acid	9.9/9.10	Prophylaxis in chickens
	Buquinolate	Quinolone	9.5	Prophylaxis in chickens
	Monensin	Antibiotic	9.8	Prophylaxis in chickens
	Clopidol	Pyridone	9.21	Prophylaxis in chickens
	Robenidine	Robenidine	9.4	Prophylaxis in chickens
	Diaveridine/sulphaquinoxaline	2,4-Diaminopyrimidine/sulphonamide	9.14/9.10	Prophylaxis in chickens
	Decoquinate	Quinolone	9.5	Prophylaxis in chickens
	Nicarbazin	Carbanilide–pyrimidine complex	9.21	Prophylaxis in chickens

	Drug	Chemical class	Section	Application
	Ormetoprim/sulphadimethoxine	2,4-Diaminopyrimidine/sulphonamide	9.14/9.10	Prophylaxis in chickens
	Methyl benzoquate	Quinolone	9.5	Prophylaxis in chickens
	Zoalene	Nitrobenzamide	9.21	Prophylaxis in chickens
	Amprolium/ethopabate/sulphaquinoxaline	Thiamine analogue/2-substituted p-aminobenzoic acid/sulphonamide	9.9/9.10/9.10	Prophylaxis in chickens
Toxoplasmosis	Pyrimethamine/dapsone	2,4-Diaminopyrimidine/sulphone	9.14/9.10	Treatment in man
Malaria	Quinine	—	9.16	Treatment in man
	Primaquine	8-Aminoquinoline	9.7	Post-curative treatment in man
	Chloroquine	4-Aminoquinoline	9.16	Prophylaxis/treatment in man
	Pyrimethamine	2,4-Diaminopyrimidine	9.14	Prophylaxis in man
	Pyrimethamine/dapsone	2,4-Diaminopyrimidine/sulphone	9.14/9.10	Prophylaxis in man
	Minocycline	Tetracycline	9.20	Treatment in man
	Clindamycin	Chlorinated lincomycin	9.20	Treatment in man
	WR 30090	Quinoline–methanol	9.16	Treatment in man
	WR 33063	Phenanthrene–methanol	9.16	Treatment in man
Babesiasis	Berenil	Diamidine	9.18	Treatment in cattle
Theileriasis	Menoctone	Napthoquinone	9.6	Trials in cattle in progress for east coast fever

* These are all in chapter 9 where biochemical modes of action of drugs are discussed.

tetracyclines, such as minocycline, and chlorinated lincomycins, such as clindamycin, are sometimes used. Two new drugs, WR 30090 and WR 33063, have recently been through successful clinical trials but have not yet come into general use.

2.3.9 Babesiasis

Originally, drugs such as phenanidine, quinuronium sulphate and amicarbalide were used, but recently these have been superseded by the antitrypanosomal drug, berenil. Imidocarb is also quite active, but has been withdrawn, temporarily at least, because of problems with tissue residues.

2.3.10 Theileriasis

There is at present no drug on general release for use against this disease, but recently menoctone has been undergoing veterinary trials and the results so far are very encouraging.

2.4 General Conclusions

Lack of commercially available vaccines means that, for the present, control of most of the parasitic diseases of man and his domestic animals relies largely on drugs. This is unfortunate since, although drugs have now been developed for most of the diseases, they are not always freely available (for example, suramin), many cannot be given orally (for example, pentamidine), they do not always work against all stages of the disease (for example, suramin, lampit, chloroquine), they are often highly toxic (for example, Mel B) and resistance to them is becoming an increasing problem (for example, ethidium, antrycide, chloroquine and all the anticoccidial drugs). There is clearly, therefore, the need for a new generation of drugs.

Most of the present drugs were developed empirically, that is, compounds for screening were synthesised or selected at random. However, a more rational approach is possible. There are many reasons of a purely academic nature to justify the biochemical studies of parasitic protozoa which will be described in this book. Most, however, have been carried out, at least in part, in the hope that if a reservoir of information could be built up about the metabolism of the parasites and the relationship of this to the metabolism of the host, it might be possible to develop a new generation of drugs much more rationally. Thus, in the following chapters, particular emphasis will be placed on biochemical differences between parasite and host, which have been discovered and might be exploited in the formulation of new chemotherapeutic agents. In the final chapter, an assessment will be made of just how far we have advanced along the road towards truly rational chemotherapy of parasitic protozoan diseases.

2.5 Further Reading

Cohen, S. and Sadun, E. (eds) (1976). *Immunology of Parasitic Infections*. Blackwell Scientific Publications, Oxford

Gutteridge, W. E. (1976). Chemotherapy of Chagas' disease: the present position. *Trop. Dis. Bull.*, **73**, 699–705

Peters, W. (1974). Recent advances in antimalarial chemotherapy and drug resistance. *Adv. Parasit.*, **12**, 69–114

Peters, W. (1976). The search for antileishmanial agents. In *Biochemistry of Parasites and Host–Parasite Relationships* (ed. H. Van den Bossche), Elsevier-North Holland Biomedical Press, Amsterdam, pp. 523–535

Roitt, I. (1974). *Essential Immunology*, second edition. Blackwell Scientific Publications, Oxford

Ryley, J. F. and Betts, M. J. (1973). Chemotherapy of chicken coccidiosis. *Adv. Pharmac. Chemother.*, **11**, 221–293

Steck, E. A. (1971). *The Chemotherapy of Protozoan Diseases*. Division of Medical Chemistry, Walter Reed Army Institute of Research, Washington D. C.

Williamson, J. (1976). Chemotherapy of African Trypanosomiasis. *Trop. Dis. Bull.*, **73**, 531–542

3 Methodology

3.1 Introduction

Biochemical investigation of parasitic protozoa usually can be carried out with a minimum of modifications of standard techniques. Methods requiring only small amounts of material, such as those involving radiotracers, are often favoured. The preparation of the protozoan material in a form suitable for the application of the standard techniques, however, requires the specialist methodology which will be outlined in this chapter. The reader is urged at this point to resist the temptation to turn to chapter 4 since it is important that the problems and present limitations of the methodology for biochemical work with parasitic protozoa should be clearly understood.

3.2 Laboratory Maintenance

Biochemical studies of parasitic protozoa can most easily be carried out with organisms collected from axenic cultures of the species which cause the disease in man or domestic animals since large quantities of uncontaminated material are then readily available. Only two, *Trichomonas* and *Entamoeba*, of the nine genera under consideration, however, can be handled in this way. The bloodstream forms of *Trypanosoma brucei* and *T. congolense* and the intraerythrocytic stages of *Plasmodium* now can be grown serially *in vitro*. The systems are complex, however, and involve low parasitaemias (appendix B) so that their potential for use in biochemical studies is questionable and, as yet, untested. With the other parasitic protozoa, either organisms can only be cultured *in vitro* axenically in a form which is different morphologically and in some cases biochemically from that which occurs in the vertebrate host (most *Trypanosoma* species, *Leishmania*) or they cannot be cultured serially at all (*Eimeria, Toxoplasma, Babesia, Theileria*). Thus, with the exceptions of *Trichomonas* and *Entamoeba*, parasites are grown in laboratory animals or occasionally, now, in tissue culture (*Trypanosoma cruzi, Toxoplasma, Eimeria, Theileria*) and subsequently isolated from the appropriate tissue or organ. Studies with forms morphologically different from the stages parasitic to the vertebrate host are carried out also (*Trypanosoma, Leishmania*) since they can be cultured axenically and are considered similar to stages which occur in insect vectors and which cannot be isolated readily. Such forms have been known generally as culture forms, an inexact term which must be used with care.

Until recently, even these stages have not been easy to culture and therefore a considerable amount of work has been done using, for example, members of the genus *Crithidia* (especially *C. fasciculata, C. luciliae* and *C. oncopelti*) as model trypanosomes (but see appendix C).

In most studies, it is possible to use the species of protozoan which actually causes the disease (for example *Trypanosoma, Leishmania, Trichomonas, Entamoeba, Eimeria, Toxoplasma* and *Theileria*). *Plasmodium* species, however, are highly host specific so that in most cases species (for example, *P. knowlesi* and *P. berghei*) other than the human pathogens have to be used. This situation may change now that *Plasmodium* can be cultured serially (given the limitations above). Related species may be used also to avoid problems of working with human pathogens (for example *Trypanosoma brucei* for *T. rhodesiense* and *T. gambiense*; *Entamoeba invadens* for *E. histolytica*).

The information in this section is summarised in table 3.1.

Table 3.1 Source of protozoa used for biochemical investigation

Genus	In vitro				In vivo	
	Actual species with correct morphology	Actual species with altered morphology	Model species	Actual species tissue culture	Actual species	Model species
Trypanosoma	−	±	+	−	+	+
T. cruzi	−	+	±	±	+	±
Leishmania	−	+	−	−	±	−
Trichomonas	+	−	±	−	−	−
Entamoeba	+	−	±	−	−	−
Eimeria	−	−	−	±	+	−
Toxoplasma	−	−	−	−	+	−
Plasmodium	−	−	−	−	±	+
Babesia	−	−	−	−	−	+
Theileria	−	−	−	+	−	−

−, not used so far; ±, used only occasionally; +, used most frequently

3.3 Maintenance *in vivo*

The laboratory host of choice is usually the rat since it is of reasonable size, easy to breed and thus inexpensive to buy and generally freely available. Not all species of parasitic protozoa, however, will infect it with parasitaemias high enough to be useful for biochemical study. In these instances, other species such as mice, hamsters, cotton rats, chinchillas, monkeys and even chickens and ducks are used. Information is detailed in table 3.2.

With all genera except *Eimeria*, parasites are maintained in laboratory animals by the technique known as syringe passage. This involves the transfer

Table 3.2 Laboratory animals used as hosts for biochemical studies of parasitic protozoa

Genus	Species	Hosts
Trypanosoma	rhodesiense	Rat
	gambiense	Rat
	brucei	Rat
	evansi	Rat
	equinum	Rat
	equiperdum	Rat
	congolense	Mouse (rat-adapted strains frequently used)
	vivax	Mouse (rat-adapted strains frequently used)
	lewisi	Rat
	cruzi*	Mouse, irradiated rat, chinchilla
Leishmania	donovani*	Rodents
	tropica*	Rodents
	braziliensis*	Rodents
	mexicana*	Rodents
Eimeria	spp.*	Chicken
Toxoplasma	gondii*	Cotton rat, mouse
Plasmodium	falciparum	Aotus monkey
	vivax	Aotus monkey
	knowlesi	Rhesus monkey
	berghei	Rat
	vinckei	Mouse
	chabaudi	Mouse (rat-adapted strain exists)
	gallinaceum	Chicken
	lophurae	Duck
Babesia	rodhaini	Rat
Theileria	parva	Cow

* Tissue culture also possible but little used so far.

of blood (*Trypanosoma, Plasmodium, Babesia, Theileria*), spleen cells or cells from a cutaneous lesion (*Leishmania*), or peritoneal macrophages (*Toxoplasma*) from an infected to an uninfected animal, usually with the aid of a syringe. Chickens are infected with *Eimeria* by feeding them oocysts obtained from the faeces of infected animals.

Species of the genera *Trypanosoma, Leishmania* and *Plasmodium*, which are normally transmitted by insect vectors, can be maintained indefinitely in laboratory animals by syringe passage. However, in many cases, the characteristics of the strain alter. In particular, most salivarian trypanosomal strains become monomorphic (long slender forms only being present—see section 1.4) and most malaria strains no longer produce gametocytes. One solution to this problem is to transmit periodically the strain through the insect vector. An alternative is the establishment of a 'stabilate'. This involves storing material from the original strain in sealed tubes in liquid nitrogen and using it to start a new line of infection after the original line has been passaged serially for a period of time.

Initiation of infection in laboratory animals is rarely done with a single organism and therefore there is no guarantee that the genotype of all the individuals in the population is the same. Variation of genotype can be a problem for the biochemist in some situations, such as work on isozymes (subsection 7.5.8) or variant-specific surface antigens (subsection 7.5.2). The problem can be overcome by the process of cloning which involves the initiation of infection in an animal or of a culture by a single organism.

The use of the cloning technique in conjunction with the establishment of a stabilate is the best way to prevent problems of irreproducibility of results occurring due to either the selection of a particular population of organisms or the accumulation of mutations in the population during syringe passage.

3.4 Isolation from Infected Animals

Some at least of the vertebrate stages of five of the seven genera of parasitic protozoa maintained routinely in laboratory animals can now be isolated free from host cellular material (the exceptions are *Babesia* and *Theileria*). However, detailed methodology has been published only for three of these (*Trypanosoma, Leishmania* and *Plasmodium*). Most non-vertebrate stages cannot be isolated in quantities sufficient for biochemical study. Thus, there is still much research to be done in this area.

The physical bases exploited in these separations are differences in density (for example *Plasmodium*) and surface charge (for example *Trypanosoma*). Density is used in techniques involving differential centrifugation (sometimes combined with the use of sucrose gradients to magnify these differences). Surface charge is exploited in techniques involving the use of the anion exchanger, DEAE cellulose, where the more negatively-charged host blood cells are retained and the less negatively-charged protozoa are eluted. Detail of the methodology for each genus is given in appendix A.

3.5 Maintenance *in vitro*

Undefined media are used most frequently for parasitic protozoa since they are usually relatively quick and cheap to prepare and defined media have not been developed for many species. Some defined media, however, are now available for *Trypanosoma, Crithidia, Leishmania* and *Trichomonas* but, with the exception of those for *Crithidia* species, they are not minimal media so that their compositions do not allow conclusions to be drawn about the biosynthetic capabilities of the parasites. Most media are prepared at near physiological pH but since most protozoa produce acids this generally decreases during growth. Antibacterial antibiotics (especially penicillin G and streptomycin) are often added to media to help prevent contamination by bacteria many of which can grow at about ten times the rate of the protozoa (protozoan generation times are mostly in the range 5–48 h). Organisms are normally grown in batch culture; little work has been done on continuous culture. Details of media for the individual genera of parasitic protozoa are given in appendix B.

3.6 Harvesting *in vitro* Cultures

As with other microorganisms, cultures of parasitic protozoa for biochemical studies are normally harvested in mid-logarithmic phase since the specific activities of enzymes and the general metabolic activity is usually much lower and more variable in organisms from stationary phase culture. Harvesting is by centrifugation, although the force and time required to pellet the cells varies with their size. Many different washing agents are used but most contain NaCl, a buffer and glucose.

3.7 Washed Cell Suspensions

The biochemical manipulations of parasites once they have been isolated from infected animals or harvested from cultures are essentially the same. For experiments involving the incubation *in vitro* of intact organisms, washed cell suspensions are prepared in media ranging from glucose-containing buffered salines to the complex defined and undefined media described in appendix B. Most species will not increase in numbers in these conditions but they are usually metabolically active for a sufficient period of time for biochemical studies to be made. However, results obtained from cells not growing normally should be treated with some caution. Included here is *Plasmodium*; this genus could not be cultured serially until recently but dilution media for infected blood, in which parasites can be maintained through two or three intraerythyrocytic cycles, have been available for a number of years.

3.8 Cell Breakage and Subcellular Fractionation

As with other microorganisms, there is no ideal way of disrupting parasitic protozoa before subcellular fractionation or enzymatic analysis. All the standard methods work (for example, freezing and thawing, alumina grinding, French press, Mickle, nitrogen cavitation apparatus) and one will be optimum for each application. Ideally, although this is rarely done, in each new piece of work all should be tested and the best selected.

Relatively little investigation has been done on techniques for the subcellular fractionation of parasitic protozoa. They have the eukaryotic type of subcellular organisation (section 1.1) and therefore the centrifugal system standard in work with mammalian cells (figure 3.1) is used frequently, at least during the initial phase of an investigation. The fractions obtained by such methods are, not surprisingly, very heterogeneous and therefore, unless further purification has been carried out, care must be taken in the interpretation of data obtained with them.

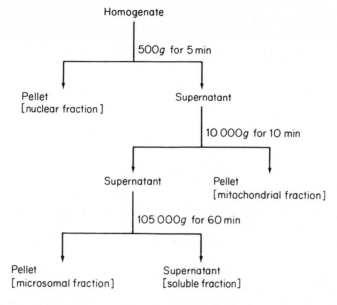

Figure 3.1 Basic scheme for the subcellular fractionation of eukaryotic cells.

3.9 General Conclusions

There are still major problems of methodology for the biochemist wishing to study parasitic protozoa. These stem from an inability to culture serially most species and thus the necessity of isolating material from infected animals. This constraint leads to high costs, lengthy procedures before biochemical studies can begin, only small amounts of material being available and the danger of artifacts due to the presence of contaminating host material in preparations. The difficulty of obtaining material is probably the most important factor limiting biochemical investigations of most parasitic protozoa and should, therefore, be borne in mind during the reading of the remainder of this book.

A more detailed coverage of two aspects of protozoan methodology, isolation and culture, is presented in appendixes A and B at the end of the book. Each includes a section on further reading.

3.10 Further Reading

Williams, B. L. and Wilson, K. (eds) (1975). *A Biologists Guide to Principles and Techniques of Practical Biochemistry*. Edward Arnold, London

4 Catabolism and the Generation of Energy. I Stages Parasitic in Vertebrates

4.1 Introduction

The production of energy in a usable form, usually adenosine triphosphate (ATP), is a basic requirement of all organisms. This energy is essential for maintaining the integrity of the organism, as well as for locomotion, growth and reproduction.

The stages in the life cycle of parasitic protozoa that are of prime importance to man are those which live in the vertebrate hosts. Thus, it is on these stages that most work has been done. However, the ever present problem of the availability of material (chapter 3) has restricted investigations to the few developmental stages readily available in quantity; other stages have been largely ignored.

In this chapter we shall consider energy stores, utilisation of substrates, transport of substrates into the cell, end products, intermediary metabolism, mechanisms of energy production, maintenance of the redox state of the cells and correlation of metabolism with cell ultrastructure. First, the situation in the mammal, the host of all the parasites under consideration except *Eimeria* and some *Plasmodium* species, will be reviewed.

4.2 The Mammalian Host

4.2.1 Energy stores

Mammals can store energy in two main forms—as polysaccharide and as lipid. The storage polysaccharide is glycogen, which is a polymer of D-glucose (figure 4.1). Linear chains of glucose units are built up by $\alpha(1 \to 4)$ glycosidic linkages, but branches occur every 8–10 residues due to additional $\alpha(1 \to 6)$ linkages (figure 4.2). Glycogen differs in structure from the typical plant storage polysaccharide, starch, in being more highly branched and thus existing as a more compact molecule. In mammals, glycogen is especially abundant in liver and muscle and usually is deposited in the form of granules, 10–40 nm in diameter, in the cytoplasm of the cell. The synthesis and breakdown of glycogen is under hormonal control and modulated also by the cellular levels of ATP, ADP, AMP and glucose-6-phosphate. Lipid energy stores normally consist of triacylglycerols (subsection 8.2.2). These storage lipids usually occur as

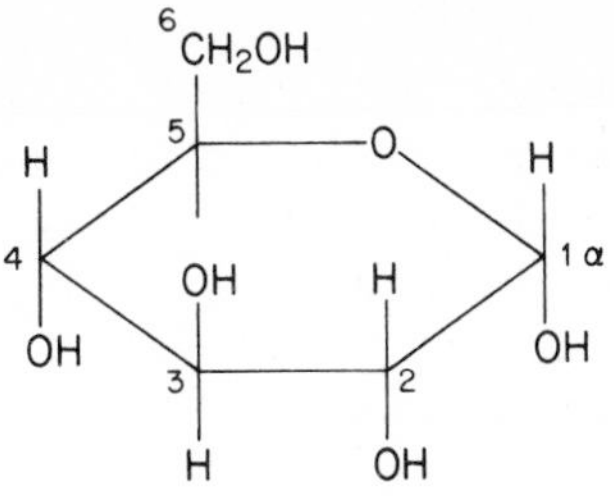

Figure 4.1 The structure of α-D-glucose.

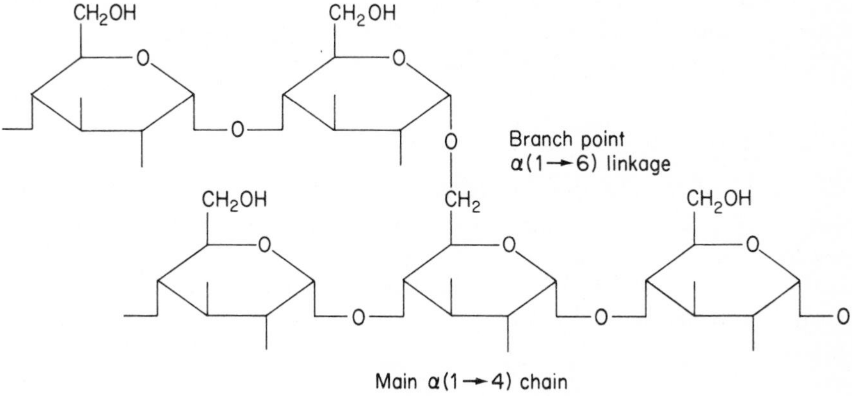

Figure 4.2 The structure of part of a glycogen molecule demonstrating $\alpha(1 \to 4)$ and $\alpha(1 \to 6)$ glycosidic linkages.

fat droplets and in mammals are especially common in adipose tissue. Synthesis of fatty acids occurs in the cytoplasm of cells, whereas catabolism occurs in mitochondria. Lipid metabolism is considered in more detail in chapter 8. Both glycogen and triacylglycerols normally are synthesised in times of abundant energy availability. Their usefulness lies in the fact that, at time of need, they can be catabolised to yield glucose-1-phosphate and acetyl coenzyme A (acetyl CoA) respectively, both of which can be further metabolised to produce energy.

4.2.2 Substrate utilisation

In mammals, macromolecules ingested as food are hydrolysed extracellularly, mainly in the stomach and intestine. Various hydrolases are involved including proteases, peptidases, lipases and enzymes which are specific to carbohydrates. The small molecules produced are absorbed through the mucosa and pass into the circulatory system.

Most mammalian cells are capable of hydrolysing macromolecules. Storage glycogen and lipid are hydrolysed by glycogen phosphorylase and lipases, respectively. Other hydrolytic enzymes, including nucleases, proteases, phosphatases and enzymes acting on polysaccharides and muco-polysaccharides, are localised in cell organelles called lysosomes. A lysosomal system is present in most cells, but it is particularly active in liver, kidney and various white blood cells. Lysosomes are known to be involved in several sub-cellular phenomena, but their activities are all basically related to the process of intracellular digestion.

The most important substrate for energy metabolism in most mammalian cells is D-glucose (figure 4.1) which is provided constantly and in abundance in the blood. Other monosaccharides and small carbohydrates are used to a minor extent. In times of need, other cellular constituents may be utilised. Glycogen and lipid stores are catabolised first, but proteins and amino acids, after deamination, can also be utilised.

4.2.3 Transport of substrates into the cell

The blood stream transports substrates to the cells. Entry mechanisms into cells are variable. Several are recognised, including simple diffusion down a concentration gradient, facilitated diffusion involving a carrier molecule and active transport which involves both a carrier molecule and the expenditure of energy, and enables substrates to be transported against a concentration gradient.

4.2.4 End products of catabolism

Most mammalian cells break down glucose aerobically to carbon dioxide and water. However, muscle cells, in .periods of high activity, will respire anaerobically and produce large quantities of lactate. This is transported to the liver and converted back to glucose (or glycogen).

4.2.5 Intermediary metabolism

The pathways involved in the catabolism of glucose to CO_2 and water are outlined in figure 4.3 and table 4.1. Only the key intermediates and points of special relevance to parasitic protozoan metabolism are given in the diagram; more detail can be obtained from any standard biochemical text. The catabolic pathways for amino acids and fatty acids are considered in chapters 7 and 8, respectively. In a typical mammalian cell, glucose is catabolised through glycolysis to pyruvate by enzymes present in the cytoplasm. Pyruvate enters the tricarboxylic acid (TCA) cycle which is localised in the mitochondria and is the common central pathway for the degradation of the two-carbon acetyl residues derived not only from carbohydrates but also from fatty acids and some amino acids. The cycle of reactions (figure 4.4), catabolised by a multi-enzyme system, splits the acetyl residues to CO_2 and releases electrons. These

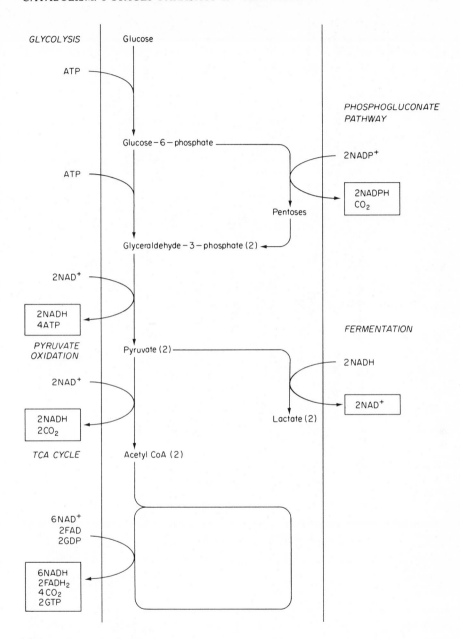

Figure 4.3 The catabolism of one molecule of glucose. The inputs and outputs (in boxes) of the various pathways are shown in the margins.

Table 4.1 Pathways of catabolism of glucose

Pathway	Overall reaction	O_2 requirement	Localisation
Glycolysis	glucose + 2ADP + 2P$_i$ + 2NAD$^+$ → 2 pyruvate + 2ATP + 2NADH + 2H$^+$	—	Cytoplasm
Phosphogluconate pathway	glucose-6-phosphate + 12NADP$^+$ + 7H$_2$O → 6CO$_2$ + 12NADPH + 12H$^+$ + P$_i$	—	Cytoplasm
Fermentation	pyruvate + NADH + H$^+$ → lactate + NAD$^+$	—	Cytoplasm
Pyruvate oxidation and TCA cycle	pyruvate + 4NAD(P)$^+$ + FAD + GDP + P$_i$ → 3CO$_2$ + 4NAD(P)H + FADH$_2$ + GTP + 4H$^+$	—	Mitochondria
Respiratory chain coupled to respiratory chain phosphorylation	NADH + H$^+$ + $\frac{1}{2}$O$_2$ + 3ADP + 3P$_i$ → NAD$^+$ + H$_2$O + 3ATP	+	Mitochondria
	FADH$_2$ + $\frac{1}{2}$O$_2$ + 2ADP + 2P$_i$ → FAD + H$_2$O + 2ATP	+	Mitochondria

are carried, via the cytochromes of the respiratory chain, to molecular oxygen, which is reduced to form water. The passage of the electrons down the respiratory chain is coupled to ATP production a process called respiratory chain phosphorylation. The structure of the respiratory chain, together with the sites of action of respiratory inhibitors and the probable sites of energy production, are

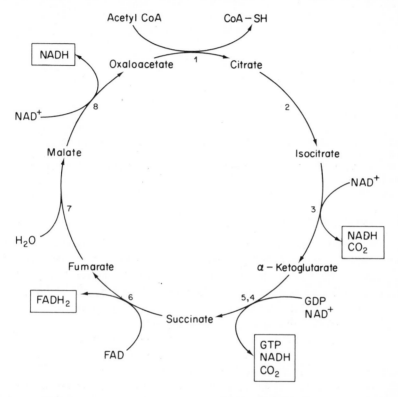

Figure 4.4 Simplified diagram of the tricarboxylic acid cycle. The end products are in boxes. Enzymes: 1, citrate synthase; 2, aconitase; 3, isocitrate dehydrogenase; 4, α-ketoglutarate dehydrogenase; 5, succinyl CoA synthetase; 6, succinate dehydrogenase; 7, fumarate hydratase; 8, malate dehydrogenase.

shown in figure 4.5. The equation for the complete oxidation of glucose as described above is

$$\text{glucose} + 6O_2 \longrightarrow 6CO_2 + 6H_2O.$$

The phosphogluconate pathway, which is not the main pathway for obtaining energy from the oxidation of glucose in animal tissues, has two major functions. First, it generates reducing power in the form of NADPH which is required in many biosynthetic reactions. Secondly, it serves to convert hexoses to pentoses which are required in the synthesis of nucleic acids.

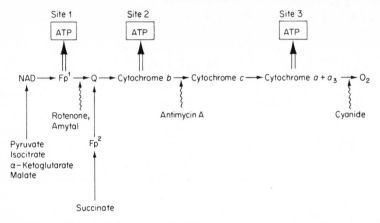

Figure 4.5 A simplified diagram of the respiratory chain in mammalian mitochondria. The approximate sites of energy production, points of action of electron transport inhibitors and places of entry of electrons from various substrates are shown. Fp, flavoprotein; Q, ubiquinone (coenzyme Q).

4.2.6 Energy production

Organisms need to produce energy in a form that is readily available for use in many cellular activities. Most organisms use ATP as the energy-rich intermediate to drive biosynthetic and other energy-requiring reactions. In these reactions, ATP transfers either an orthophosphate or a pyrophosphate group to form ADP or AMP respectively. These compounds are rephosphorylated back to ATP in energy producing reactions.

ATP is produced by mammalian cells both in substrate level phosphorylations such as occur in glycolysis and in respiratory chain phosphorylation. Mammalian cells normally oxidise glucose completely to CO_2 and water with a concomitant production, for each glucose molecule utilised, of 36 molecules of ATP. This compares with two molecules produced if glucose is converted to lactate.

4.2.7 Reoxidation of reduced coenzymes

In glycolysis, there is a net production of two molecules of reduced nicotinamide adenine dinucleotide (NADH) for each glucose molecule catabolised. For the continued survival of the cell, it is essential that this NADH is reoxidised for further participation in glycolysis. Mammalian cells normally reoxidise NADH in the mitochondria by the passage of electrons along the respiratory chain. The reduced coenzymes generated in the TCA cycle are similarly reoxidised. However, when the oxygen supply is insufficient to meet the demands of the cell (for example in muscle during intense activity) NADH can be reoxidised by the concomitant reduction of pyruvate to lactate, a reduction not linked to energy production.

4.2.8 Correlation of catabolism with cell ultrastructure

Mammalian cell organisation was discussed in chapter 1. Different organelles of the cell are the sites for different parts of metabolism, some of which have been described earlier in this chapter. Of particular interest in the biochemistry of protozoa are the mitochondria, which are abundant in mammalian cells that are actively metabolising aerobically. The mitochondria contain plate-like or septate cristae, which increase the surface area of the inner membrane in relation to mitochondrial volume. Many of the enzymes of the TCA cycle and respiratory chain are localised there. Other mitochondria contain structurally different, tubular cristae. These are less active metabolically than mitochondria with septate cristae and are characteristic of many protozoa. In some organisms mitochondria are found with no cristae; these are considered not to possess an active TCA cycle and probably also lack a respiratory chain.

4.3 Genus *Trypanosoma*

4.3.1 Energy stores

No trypanosomatids have been shown to contain substantial polysaccharide stores, whereas the absence of such stores has been confirmed for many species. The search for lipid stores has been less exhaustive but so far none has been identified unambiguously. However, energy stores must exist in a number of species (for example, *Trypanosoma cruzi*, *Crithidia fasciculata* and *Leishmania*) since these have high endogenous rates of respiration. The identity of these stores is unknown.

One interesting possibility is that polyphosphate, a polymer composed of phosphate monomers, which occurs in a wide variety of organisms, functions as an energy store. Such molecules have been isolated from *C. fasciculata* and similar structures have been identified tentatively from electron micrographs of other trypanosomatids. Various possible functions have been suggested for polyphosphates including energy reserves (phosphagens), phosphorous stores and a role in control of orthophosphate levels. Which, if any, of these suggestions is correct is unknown. An alternative suggestion is that primitive organisms used polyphosphates or pyrophosphates as energy-rich intermediates in much the same way that mammals (in which polyphosphate has little significance) and most contemporary organisms use ATP. This possibility is especially interesting with reference to the metabolism of *Entamoeba* (section 4.6). There is also evidence that polyphosphates were important constituents of the 'primordial soup'. Polyphosphates thus are of great interest and may be important in the metabolism of trypanosomatids; unfortunately, at present, they have been little studied.

4.3.2 Substrate utilisation

Undoubtedly, glucose is by far the most important exogenous substrate used by blood-stream forms of trypanosomes *in vivo*, although other hexoses and glycerol can also be utilised by some species. The physiological substrate of the intracellular amastigote stages of *T. cruzi* is unknown.

4.3.3 Transport of substrates into the cell

The mechanisms by which substrates are transported into the cell are poorly understood. However, evidence suggests that mediated systems are involved, certainly at physiological substrate concentrations, and that more than one transport site exists. *T. equiperdum* possesses three sites, two for the transport of glycerol but only one for hexoses. In other species there appear to be at least two hexose sites. *T. gambiense* takes up glucose and mannose at one site and fructose at a second, whereas in *T. lewisi* one site is specific to glucose and a second to fructose and mannose. It is possible that amino acids may be taken up at the same sites.

4.3.4 End products of metabolism

Blood-stream trypanosomes can be divided into three groups according to their end products of aerobic glucose metabolism (table 4.2). One group, containing

Table 4.2 End products of aerobic glucose metabolism
in blood-stream trypanosomes

	CO_2	Glycerol	Ethanol	Lactate	Pyruvate	Acetate	Succinate	Citrate
T. rhodesiense	±	+	—	—	+ +	—	—	—
T. gambiense	—	+	—	—	+ +	—	—	—
T. brucei	—	+	—	—	+ +	—	—	—
T. evansi	—	+	—	—	+ +	—	—	—
T. equinum	—	+	—	—	+ +	—	—	—
T. equiperdum	—	+	—	—	+ +	—	—	—
T. vivax	+	+	—	±	+	+	—	—
T. congolense	+	+	—	—	—	+	+	—
T. cruzi	+ +	—	—	±	—	+	+	—
T. lewisi	+ +	—	—	±	—	+	+	—

—, none; ±, trace; +, small quantities; + +, substantial quantities

members of the subgenus *Trypanozoon*, of which *Trypanosoma brucei* is the most studied, produce mainly pyruvate. A second group, containing the South American trypanosomes of which *T. cruzi* is the best understood, produce mainly CO_2 but also some acetate and succinate. The third group, containing the African trypanosomes *T. vivax* and *T. congolense* of the subgenera *Duttonella* and *Nannomonas*, respectively, are seemingly intermediate, producing pyruvate, other organic acids and CO_2. These groups will be considered separately for the remainder of this section, concerning the intermediary metabolism, energy production, reoxidation of NADH and cell ultrastructure.

4.3.5 Subgenus *Trypanozoon*

In the long slender forms of this group of blood-stream trypanosomes the catabolism of glucose to pyruvate (table 4.2) involves the glycolytic pathway and is associated with the expected concomitant production of energy and reduced coenzymes

glucose + 2ADP + 2P$_i$ + 2NAD$^+$ ⟶ 2 pyruvate + 2ATP + 2NADH + 2H$^+$

The evidence available suggests that the glycolytic enzymes of these protozoa are similar to those found in mammals. Some enzymes of the phosphogluconate pathway are present: they probably function only for the synthesis of pentoses and NADPH—such a situation exists in all blood-stream trypanosomes studied. Most of the enzymes of the TCA cycle and the respiratory chain are absent and the mitochondria, although present, have very few, poorly developed tubular cristae.

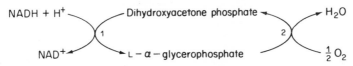

Figure 4.6 The L-α-glycerophosphate oxidase cycle present in blood-stream *Trypanosoma brucei*. Enzymes: 1, L-α-glycerophosphate dehydrogenase; 2, L-α-glycerophosphate oxidase.

Oxygen is consumed, however, although this consumption is not inhibited by cyanide indicating that it does not involve cytochrome $a + a_3$. The explanation of this oxygen utilisation became clear when the mechanism in the cell for reoxidising NADH was discovered in 1960. This system is shown schematically in figure 4.6. The combined activity of the two enzymes L-α-glycerophosphate dehydrogenase and L-α-glycerophosphate oxidase reoxidises NADH and produces water. Cyanide has no effect and there is no linked energy production. As far as is known, this system is unique to salivarian trypanosomes. The L-α-glycerophosphate dehydrogenase is possibly localised together with all the glycolytic enzymes from hexokinase to 3-phosphoglycerate kinase in microbodies and is probably not membrane bound. The enzyme from *Trypanosoma brucei* is similar to those present in *T. vivax*, *Crithidia* and *Leptomonas*, but rather different in detail from the enzymes of mammals and birds. The L-α-glycerophosphate oxidase is a particulate enzyme, possibly with several components, contains iron and copper, and is now known to be localised in the mitochondria. High oxygen tensions are required by the system (K_m approximately 2 μM).

The absence of a parallel system in mammals makes the trypanosomal L-α-glycerophosphate oxidase enzyme a prime target for rational chemotherapy.

Investigation of the enzyme system is consequently intense, although as yet the potentials of the system have not been (knowingly) exploited (but see section 9.2). The enzyme system is inhibited by various hydroxamic acids, such as *m*-chlorobenzhydroxamic acid and salicylhydroxamic acid (SHAM). However, neither of these compounds, although completely inhibiting O_2 utilisation, is lethal to the trypanosomes. This indicates that operation of the L-α-glycerophosphate oxidase is not essential to the survival of the organisms, a conclusion confirmed by the anaerobic survival of blood-stream trypanosomes. In the absence of O_2, ATP production is halved and glycerol and pyruvate are produced in equimolar amounts. A pathway such as that outlined in figure 4.7

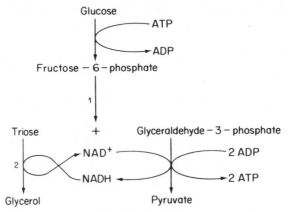

Figure 4.7 Catabolism of glucose in blood-stream *Trypanosoma brucei* in anaerobic conditions. Enzymes: 1, novel aldolase; 2, glycerol dehydrogenase.

might operate in these conditions. Whether it operates also in aerobic conditions is as yet unknown. These findings cast doubt as to whether an inhibitor of the L-α-glycerophosphate oxidase system will alone be an effective antitrypanosomal agent. Thus it is of interest to note that it has been shown recently that SHAM and glycerol together represent a potent trypanocidal combination.

The situation described above is that present in the long slender blood-stream trypanosomes. They are the dividing stages and the forms present in the acute phase of an infection. Other blood-stream forms exist, however, the short stumpy forms (see section 1.4). These occur later in an infection, mainly during remission, and are considered the only stages that are infective to the tsetse fly, that is they are responsible for transmission. The energy metabolism of the short stumpy forms is rather different from that of the long slender forms, and will be considered in more detail in the next chapter, when the insect stages of trypanosomes are discussed (section 5.2).

4.3.6 *Trypanosoma vivax* and *T. congolense*

These species will be considered together because they both seem to be intermediate between the *Trypanozoon* species and *Trypanosoma cruzi* with respect to their energy metabolism (table 4.2). Another similarity is that little is known about the metabolism of either. They both possess tubular cristae in the mitochondria and some of the enzymes in the TCA cycle are present although it is unlikely that either a fully functional TCA cycle or cytochrome chain occurs. The end products of metabolism, which include acetate and succinate, give some confirmation of this suggestion, although the mechanisms of their formation are not known. NADH oxidation has not been studied in detail but probably involves a L-α-glycerophosphate oxidase system.

4.3.7 *Trypanosoma cruzi*

T. cruzi is rather different from the trypanosomes of the subgenus *Trypano-zoon*, and also less studied (table 4.2). The available evidence for the bloodstream trypomastigotes suggests that glucose is catabolised via glycolysis to pyruvate. Again the phosphogluconate pathway probably only proceeds as far as pentose and NADPH production, although it has been claimed that in some strains at least approximately 25 per cent of glucose catabolism proceeds via this pathway rather than by glycolysis. Pyruvate can be metabolised further in four ways, in three of which NADH is reoxidised concomitantly. Some pyruvate (about 55 per cent) enters the TCA cycle and is fully oxidised to CO_2. The NADH produced is reoxidised via a cytochrome chain, which is available also to reoxidise the NADH generated in glycolysis. The reaction of oxygen with the reduced terminal electron acceptor is in part cyanide sensitive, indicating a cytochrome $a + a_3$, but other terminal oxidases probably are present as in the culture forms of *Trypanosoma brucei* (see section 5.2). The presence of mitochondria with plate-like cristae supports the existence of these pathways. A little pyruvate is excreted as such, a small amount (about 6 per cent) is reduced to lactate, by the enzyme lactate dehydrogenase, and is excreted; the remainder is metabolised to succinate (about 12 per cent) and acetate (about 17 per cent). It is not known how either is produced but the carboxylation of pyruvate or phosphoenolpyruvate (PEP) to malate or oxaloacetate and the reduction of these compounds by the TCA cycle enzymes acting in reverse, is probably the source of the succinate. Our present understanding of the catabolism of glucose by the blood-stream forms of *T. cruzi* is presented schematically in figure 4.8.

Fatty acids also are oxidised by blood-stream forms of *T. cruzi*, yielding CO_2 and presumably concomitant energy. Nothing is known of the mechanisms involved, although probably acetyl CoA is produced and this is oxidised in the TCA cycle.

Blood-stream forms of *T. cruzi* thus obtain energy from several sources including respiratory chain phosphorylation coupled to electron transport and substrate level phosphorylation both during glycolysis and possibly also during the production of succinate and acetate. Blood-stream *T. cruzi* occurs in two morphological forms. The possible analogy with the two morphological forms of blood-stream *T. brucei* is obvious, but as yet uninvestigated. The metabolism of the blood-stream forms of *T. lewisi* is probably similar to *T. cruzi*.

Much less is known about the energy metabolism of the intracellular amastigotes of *T. cruzi*. Preliminary evidence suggests that it is similar to that of the blood-stream forms, although this remains to be substantiated, particularly the identification of products. The presence of mitochondria with plate-like cristae supports this suggestion.

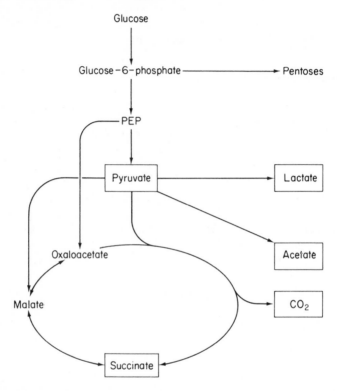

Figure 4.8 Glucose catabolism in blood-stream *Trypanosoma cruzi*. The end products are in boxes. See text for details.

4.3.8 Genus *Crithidia*

This is not a useful model for energy metabolism in blood-stream *T. brucei*—a full TCA cycle and cytochrome system are present. It is possibly a reasonable model for the culture forms of trypanosomes (see section 5.2), which are morphologically more similar.

4.4 Genus *Leishmania*

The amastigote stages of *Leishmania* have been studied very little. They possess mitochondria with a few plate-like cristae, and oxygen consumption is cyanide sensitive in part, suggesting the existence of multiple terminal oxidases.

4.5 Genus *Trichomonas*

4.5.1 Energy stores

Glycogen normally is present, distributed throughout the cell and is responsible for 10–30 per cent of its dry weight. Enzymes similar to those present in mammalian cells catalyse the synthesis and catabolism of the glycogen.

4.5.2 Substrate utilisation

Trichomonads will utilise hexoses, disaccharides and polysaccharides with $\alpha(1 \rightarrow 4)$ glycosidic linkages. Although maltose supports growth best, glucose is used most frequently in laboratory culture media. Amylases are present in trichomonads. Other enzymes demonstrated in *Trichomonas foetus* are β-glucosidases, galactosidases, mannosidases, hexosaminidases, fucosidases, neuraminidases, and N-acetylhexosaminidases. Many are apparently localised in lysosome-like organelles.

4.5.3 Transport of substrates into the cell

Little is known.

4.5.4 End products of catabolism

Both *T. vaginalis* and *T. foetus* excrete acetate, CO_2 and H_2, and, in addition, *T. vaginalis* produces lactate and malate and *T. foetus* excretes succinate.

4.5.5 Intermediary metabolism and correlation with the cell ultrastructure

Trichomonads normally are considered to be anaerobic, although they will grow under low oxygen tensions. Mitochondria are absent, as are cytochromes and a functional TCA cycle. The catabolism of glucose to pyruvate is through the glycolytic pathway, which is similar in form to that present in mammalian cells. The phosphogluconate pathway functions in the synthesis of pentoses and NADPH.

Pyruvate has various fates. In *T. vaginalis* and *T. gallinae* some is reduced to lactate, a reaction catalysed by lactate dehydrogenase, which is a regulatory enzyme in *T. gallinae*. In *T. foetus*, some pyruvate is reduced to succinate. This apparently involves a carboxylation to oxaloacetate (OAA) or malate. In mammals three mechanisms of these conversions are known

$$PEP + GDP + CO_2 \longrightarrow OAA + GTP \text{ [PEP carboxykinase (GTP)]}$$

$$pyruvate + NADPH + CO_2 \longrightarrow malate + NADP^+ \text{ [malic enzyme]}$$

$$pyruvate + ATP + CO_2 + H_2O \longrightarrow OAA + ADP + P_i \text{ [pyruvate carboxylase]}$$

A further carboxylation reaction has been found in green plants

$$PEP + CO_2 \longrightarrow OAA + P_i \quad [PEP\ carboxylase]$$

Present evidence indicates that PEP carboxykinase (GTP) is the most important enzyme in *T. foetus*, although the malic enzyme also is present. Succinate is produced from oxaloacetate or malate by reactions catalysed by the enzymes malate dehydrogenase, fumarate hydratase and fumarate reductase.

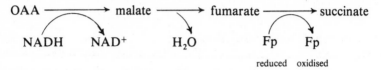

The final step in this pathway, the reduction of fumarate to succinate, in which an unknown electron carrier, probably a flavoprotein (Fp), is involved, is catalysed by the enzyme fumarate reductase. Enzymes carrying out similar conversions in some anaerobic helminths, for example *Ascaris*, have been shown to be linked to energy production, and are of considerable importance as sources of ATP for them. The enzymes in *Trichomonas* species have been little studied, but the idea of succinate production linked to a concomitant energy production is very attractive. *T. vaginalis* excretes malate without further metabolism.

A third fate of pyruvate is conversion to acetate, CO_2 and H_2. This pathway is shown schematically in figure 4.9. As yet the electron transfer molecule X is unidentified, but is probably similar to the iron–sulphur proteins ferredoxin and

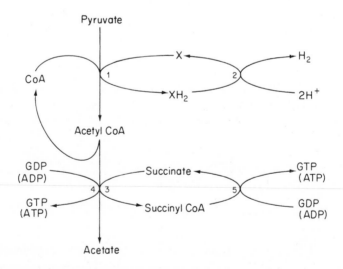

Figure 4.9 Conversion of pyruvate to acetate in trichomonads. Enzymes: 1, pyruvate: X oxidoreductase (pyruvate synthase); 2, hydrogenase; 3, succinyl acetyl CoA transferase; 4, acetyl thiokinase; 5, succinyl thiokinase. X, unknown electron transfer protein.

flavodoxin which perform similar functions in some anaerobic bacteria, including *Clostridium*. The acetyl thiokinase and succinyl thiokinase activities are probably both expressions of the same enzyme. In both cases ADP or GDP are equally good acceptors. The enzymes responsible for this cleavage of pyruvate are all localised in distinct cellular organelles, morphologically similar to microbodies, named hydrogenosomes, which are as yet unique to *Trichomonas*. These organelles contain cardiolipin, superoxide dismutase similar to that found in bacteria and mitochondria, and circular DNA of 2–3 μm contour length. The possibility that they are derived from bacteria similar to *Clostridium* is attractive.

The hydrogenosomes, and the metabolic systems present within them, are unparalleled in the mammalian cell, and their chemotherapeutic potential is obvious. The only clinically useful antitrichomonal drugs available at present, the 5-nitroimidazoles, in fact probably owe their activity to these hydrogenosomes (section 9.5) although the discovery of their activity predates that of the hydrogenosomes by about 15 years.

4.5.6 Energy production

The sources of energy for *Trichomonas* species are not all fully understood. Clearly energy is produced during the glycolytic sequence and energy production is linked to the formation of acetate. The possibility of energy-linked succinate production was described above.

When oxygen is present, as is probably the situation *in vivo*, the energy production increases from 5ATP/glucose, in anaerobic conditions, to 7ATP/glucose. Concomitantly more acetate, less succinate and no H_2 are produced. This is probably a consequence of the high affinity of O_2 for the reduced carrier X, which results in cessation of H_2 production and diversion of more pyruvate into the acetate pathway. These results also show that the formation of acetate is linked to a greater energy production than is the formation of succinate. Whether this O_2 effect is of physiological significance is unknown. Certainly under low oxygen tensions the hydrogenosomes are kept anaerobic by a cytoplasmic NADH oxidase which reduces the O_2 entering the cell.

4.5.7 Reoxidation of reduced coenzymes

The reoxidation of the NADH produced during glycolysis occurs in the production of lactate and succinate. Whether the metabolic pathways in the hydrogenosomes also play a part is unknown, but possibly the electron transfer protein X can be reduced by NADH directly. When oxygen is present some NADH is reoxidised by the cytoplasmic NADH oxidase.

A schematic representation of our present understanding of the catabolism of glucose and production of energy in trichomonads is given in figure 4.10.

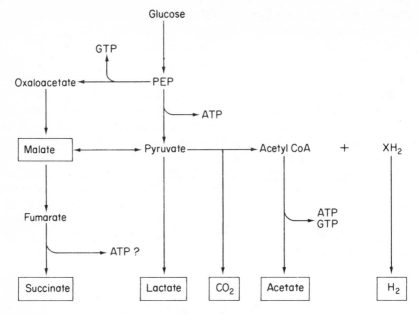

Figure 4.10 Schematic representation of the catabolism of glucose in trichomonads. The end products are in boxes. See text for details.

4.6 Genus Entamoeba

4.6.1 Energy stores

The trophozoites of *Entamoeba histolytica* contain large polysaccharide stores, mainly glycogen, which are spread throughout the cell.

4.6.2 Substrate utilisation

Very few carbohydrates will support the growth of *E. histolytica*. Some strains are specific to D-glucose or a polymer of D-glucose alone. Others will utilise also D-galactose. Amylases are present.

This essential requirement for specific carbohydrates may explain the invasion of the tissues by *E. histolytica*; a deficiency of available substrate in the lumen of the intestine causing the migration of the protozoan to the car-·bohydrate rich tissues of the wall.

4.6.3 Transport of substrates into the cell

Few studies of glucose transport have been made, although there is evidence that this is the rate-limiting step of glycolysis. There is a specific transport system for glucose. This contrasts with non-parasitic amoebae which apparently use pinocytosis alone. Phlorrhizin, a general inhibitor of sugar transport in animal tissues, has no effect in *E. histolytica*, indicating differences in the transport systems.

4.6.4 End products of glucose catabolism

The end products of the catabolism of glucose by *E. histolytica* in anaerobic and aerobic conditions are given in table 4.3. Oxygen causes a cessation of H_2 production and an increased acetate production, so that the ratio of acetate to ethanol excreted increases from 0.37 in anaerobic conditions to 2.2 in aerobic conditions. Very recent work indicates that under anaerobic conditions some strains of axenic *E. histolytica* produce neither H_2 nor acetate; however in the presence of O_2 acetate is formed.

Table 4.3 Catabolism of glucose by *Entamoeba histolytica*

	Anaerobic	Aerobic
Glucose utilised (μmol/min/ml packed cells*)	0.7	0.4
O_2 utilised (μmol/min/ml packed cells*)	—	0.6
Products (mol/mol glucose utilised)		
CO_2	2.0	2.1
H_2	0.6	0
Acetate	0.5	1.5
Ethanol	1.4	0.7

* 1 ml packed cells contain approximately 1.7×10^8 organisms

4.6.5 Intermediary metabolism

Investigations of the catabolism of glucose by *E. histolytica* have revealed many unusual enzyme systems which, together with the absence of mitochondria, a TCA cycle and cytochromes, make this aspect of the metabolism quite distinct from that of mammals.

Some strains of *E. histolytica* will utilise only glucose and no other monosaccharides will support growth. This specificity seems to be at the level of uptake into the cell, for the hexokinase which catalyses the reaction

$$\text{D-glucose} + \text{ATP} \longrightarrow \text{D-glucose-6-phosphate} + \text{ADP}$$

can phosphorylate many hexoses in cell-free systems. Unusually, the hexokinase is under regulation, high levels of AMP inhibiting its activity. The significance of this control is obscure, and possibly other modulators will be discovered in the future. In most organisms the conversion of fructose-6-phosphate to fructose-1,6-diphosphate is catalysed by the enzyme phosphofructokinase (PFK). In mammals a single enzyme exists, but *E. histolytica* possesses two. One is similar to the mammalian enzyme in catalysing the reaction

$$\text{D-fructose-6-phosphate} + \text{ATP} \longrightarrow \text{D-fructose-1,6-diphosphate} + \text{ADP}$$

However, unlike the mammalian enzyme it is not under regulation. The second enzyme is so far unique to *E. histolytica* in being specific to pyrophosphate (PP_i) rather than ATP. It has been named PFK (PP_i) and probably is the more important enzyme of the two. It is not a regulatory enzyme. Control of mammalian glycolysis is largely at the level of PFK which is inhibited by high concentrations of ATP, citrate and long chain fatty acids, but is stimulated by ADP or AMP. This is the first enzyme common to glycolysis irrespective of the carbohydrate used. However, no control of the amoebal PFKs has been reported yet. One possible explanation is that as only glucose is used, a better control point for glycolysis is hexokinase or even the transport of glucose into the cell. The phosphoglycerate kinase of *E. hisolytica* also differs from the mammalian enzyme. It is specific for GTP rather than ATP

3-phosphoglyceroyl phosphate + GDP   3-phosphoglycerate + GTP

In mammals PEP is converted mainly to pyruvate in a reaction catalysed by pyruvate kinase; this enzyme is absent from *E. histolytica*. Instead it possesses the enzyme pyruvate phosphate dikinase which has a specificity for pyrophosphate

$$PEP + AMP + PP_i \longrightarrow pyruvate + ATP + P_i$$

In the amoebae some PEP is carboxylated to oxaloacetate in a reaction catalysed by PEP carboxytransphosphorylase, another enzyme with a specificity for pyrophosphate.

$$PEP + CO_2 + P_i \longrightarrow OAA + PP_i$$

The oxaloacetate produced can be converted to pyruvate by the action of the two enzymes, malate dehydrogenase (specific to NAD) and the malic enzyme (specific to NADP). Thus a branched pathway between PEP and pyruvate exists. The advantage of such a branch is obscure, although clearly the four-carbon intermediates may be useful in biosynthetic reactions and the NADPH could be used in the synthesis of fatty acids.

Pyruvate has several possible fates. Ethanol is produced by the decarboxylation of pyruvate to acetaldehyde and a subsequent reduction catalysed by alcohol dehydrogenase. This second enzyme is, unusually, specific to NADPH in *E. histolytica*. The enzyme in *E. invadens* is, however, specific to NADH. Alternatively, pyruvate is converted via acetyl CoA and acetyl phosphate to acetate. The mechanism of these conversions seems to depend on the environmental conditions. In the presence of oxygen, the enzyme pyruvate oxidase (CoA acetylating) is active

$$pyruvate + CoA + O_2 \longrightarrow acetyl\ CoA + CO_2 + H_2O_2$$

FAD apparently is required, although its exact role is uncertain. However, in anaerobic conditions pyruvate is converted to acetyl CoA, probably in a reaction similar to that found in *Trichomonas* (section 4.6). Iron–sulphur proteins are present in *Entamoeba* and probably serve as the electron transport protein

X in this reaction sequence. However, no organelles similar to hydrogenosomes are present in *E. histolytica*. The presence of such a metabolic pathway would explain both the appearance of gaseous hydrogen as an end product and the sensitivity of organisms to 5-nitroimidazoles (section 9.5). Different mechanisms of converting acetyl CoA to acetate are found in cells. Most higher organisms use a single enzyme, acetyl thiokinase

$$\text{acetyl CoA} + \text{ADP} + P_i \longrightarrow \text{acetate} + \text{CoA} + \text{ATP}$$

However, other mechanisms are known

$$\text{acetyl CoA} + \text{AMP} + PP_i \longrightarrow \text{acetate} + \text{CoA} + \text{ATP} \quad [\text{acetyl CoA synthetase}]$$

$$\text{acetyl CoA} + \text{carnitine} \longrightarrow \text{acetyl carnitine} + \text{CoA} \quad [\text{carnitine acyltransferase}]$$

$$\text{acetyl carnitine} \longrightarrow \text{acetate} + \text{carnitine} \, [\text{acetyl carnitine hydrolase}]$$

$$\text{acetyl CoA} + \text{succinate} \longrightarrow \text{succinyl CoA} + \text{acetate} \, [\text{succinyl acetyl CoA transferase}]$$

$$\text{succinyl CoA} + \text{ADP (GDP)} + P_i \longrightarrow \text{succinate} + \text{CoA} + \text{ATP (GTP)} \quad [\text{succinyl thiokinase}]$$

Many microorganisms also require two enzymes working in conjunction, phosphate acyltransferase and ATP-dependent acetate kinase

$$\text{acetyl CoA} + \text{phosphate} \longrightarrow \text{acetyl phosphate} + \text{CoA}$$

$$\text{acetyl phosphate} + \text{ADP} \longrightarrow \text{acetate} + \text{ATP}$$

E. histolytica possesses a pathway similar to that of many bacteria, although the enzymes differ in detail. In fact the acetate kinase of one, non-pathogenic, strain of *E. histolytica* is similar to the bacterial enzyme. However, the acetate kinase from another strain is unusual and is specific for pyrophosphate rather than ATP. Very recently an enzyme similar to, but different from, acetyl thiokinase has been found in axenic *E. histolytica*. The pathways involved in glucose catabolism of *E. histolytica* are diagrammatically represented in figure 4.11. Very recently it has been suggested that enzymes (hydrogenase, phosphate acyltransferase) previously reported to be amoebic may have been of bacterial origin. This awaits confirmation. The absence of hydrogenase would explain the lack of both H_2 production and acetate production under anaerobic conditions.

4.6.6 Energy production, NADH oxidation and discussion

It is obvious from the preceding section that the catabolism of glucose in *E. histolytica* is unusual in many respects, and still not understood. One peculiarity is the number of enzyme reactions in which pyrophosphate is in-

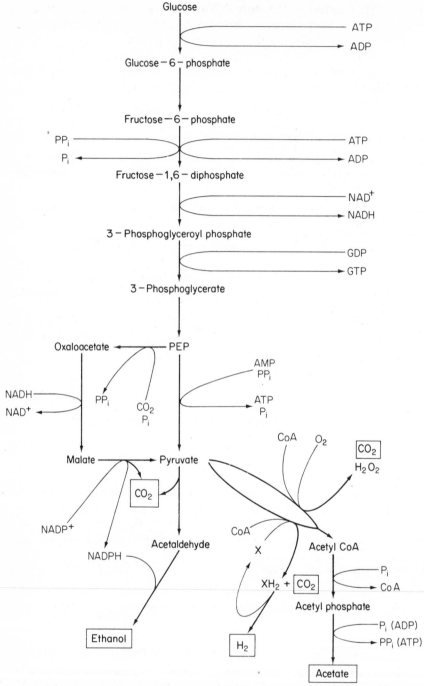

Figure 4.11 Schematic representation of the catabolism of glucose in *Entamoeba histolytica*. End products are in boxes.

volved—four have been discovered so far. Pyrophosphate usually is found in very low concentrations in mammalian cells, whereas in *E. histolytica* it is present at high concentrations. As discussed previously (subsection 4.3.1), it is thought that in primitive cells pyrophosphate may have served the same function as ATP does now in higher organisms. The abundance of pyrophosphate and of pyrophosphate-specific enzymes in *E. histolytica* is an indication that amoebal metabolism may have emerged very early from the evolutionary pattern of development followed by most other organisms investigated. ATP also is important in *E. histolytica*, however, although it will be interesting to see if this requirement can be replaced by pyrophosphate in any more reactions.

The net production of energy is one molecule of ATP for each molecule of glucose catabolised to pyruvate assuming equal flux through the two branches from PEP to pyruvate. Additional energy is produced in the catabolism to acetate. Exactly how the redox potential of the cell is maintained is, at present, difficult to understand. Reduced nucleotides are reoxidised in the synthesis of ethanol, but this cannot account for all the NADH. One possible explanation is that some NADH is reoxidised by the highly electronegative electron transfer protein that is an integral component of the anaerobic conversion of pyruvate to acetyl CoA. It was suggested that such a reaction may occur also in *Trichomonas* (section 4.5). A transhydrogenase discovered in *Entamoeba* is probably very important in the maintenance of the redox state of the cell, especially in the presence of oxygen (see below). The reaction catalysed is

$$NADH + NADP^+ \longrightarrow NAD^+ + NADPH$$

The localisation of the enzyme in *Entamoeba* is uncertain, but must differ from that in mammalian cells where it is present on the inner mitochondrial membrane. The mammalian enzyme is considered to be potentially important in energy production when acting in the formation of NADH. A similar important role for the *Entamoeba* enzyme is possible.

The role of oxygen in the metabolism of *E. histolytica* is problematical—*Entamoeba* is considered a putative anaerobe. However, there is no doubt that oxygen is consumed, with high affinity, when available. One possibility is that this is a detoxification mechanism, which could be necessary as the parasite clearly must encounter high oxygen concentrations in the blood. The discovery of the pyruvate oxidase (CoA acetylating) activity can explain both the oxygen utilisation, and the change in the end products in aerobic conditions. Such an alteration in end products is favourable to the organism because acetate production is energy linked. Oxygen utilisation can be explained also by the presence of a NADPH oxidase activity, whereby coenzyme is reoxidised by O_2. Iron–sulphur proteins are involved. The coupling of this reaction to the transhydrogenase described above would allow the reoxidisation of NADH by O_2. Such a system could be important to a facultatively aerobic organism both in the maintenance of the redox state and in the scavenging of toxic oxygen. *E. invadens* possesses a NADH oxidase which could serve the same purpose.

The existence of two enzymes metabolising pyruvate to acetyl CoA is surprising. A possibility is that in fact only one enzyme exists. In anaerobic conditions the reduced electron transfer protein (similar to ferredoxin and flavodoxin) would be reoxidised by the enzyme hydrogenase and H_2 would be evolved. However, when oxygen was present this would react with the reduced carrier directly, causing the cessation of H_2 evolution and the production of hydrogen peroxide. A greater affinity of the reduced electron transfer protein for oxygen than for the hydrogenase enzyme would explain the shift in the metabolism from ethanol to acetate production. This proposed enzyme system is very similar to the pyruvate–X oxidoreductase present in *Trichomonas* (section 4.5). It is possible also that the NADPH oxidase activity described above is another manifestation of the reaction of O_2 with this reduced electron transfer protein.

This multitude of unusual reaction mechanisms makes *E. histolytica* a good organism for rational chemotherapy. Work is in progress, but at present 5-nitroimidazoles are the drugs of choice (section 9.5).

4.7 Genus *Eimeria*

Very little is known of the metabolism of the intracellular stages of *Eimeria* species. The mature trophozoites and macrogametocytes possess a storage polysaccharide, more similar in structure to amylopectin (typical of plants) than glycogen, although there is little or none in the young trophozoites or the microgametocytes. Macrogametocytes possess a β-galactosidase but no β-glucosidase. Preliminary evidence from histological staining suggests that the young trophozoites obtain their energy requirements mainly from glycolysis, whereas the schizonts may utilise the TCA cycle and possibly a cytochrome chain. This is similar to the suggestions put forward for the blood-stream stages of *Plasmodium knowlesi* (section 4.8) but further investigations are required in both cases before the situation should be considered as proven. It is likely that the anticoccidial quinolones act by disrupting the respiratory chain (section 9.2).

4.8 Genus *Plasmodium*

In this section the vertebrate stages of both the mammalian malarias and the bird malarias, which are often used as mammalian models, will be considered. All the discussion concerns the intraerythrocytic asexual stages except when stated to the contrary.

4.8.1 Energy stores

No energy stores have been definitely identified in any of the stages of the parasites in the vertebrate hosts. There is some evidence for lipid stores in the intraerythrocytic stages of *Plasmodium lophurae*.

4.8.2 Substrate utilisation

Several monosaccharides and disaccharides can support growth *in vitro*, but probably glucose is the main physiological substrate of these blood-dwelling parasites.

4.8.3 Transport of substrates into the parasite cell

Simple sugars enter by a mediated process, and in *P. lophurae* there are two distinct loci with different sugar specificities. A problem of any intracellular parasite is to obtain nutrients across two cell membranes, its own and that of the host. *Plasmodium* has overcome this difficulty by inducing, by an unknown mechanism, the host cell to lose (partially) the ability to regulate the passage of molecules across its cell membrane. The host cell becomes freely permeable to many complex molecules which therefore can be taken up with little or no expenditure of energy. This is a considerable advantage to the intracellular parasite.

4.8.4 End products

All *Plasmodium* species have been found to produce lactate to a greater or lesser extent (table 4.4). The intermediary metabolism, energy production and methods of reoxidising NADH of primate, rodent and avian malaria will be considered separately.

Table 4.4 End products of glucose catabolism in malaria parasites

	Primate malaria (*P. knowlesi*)	Rodent malaria (*P. chabaudi*)	Avian malaria (*P. lophurae*)
Approximate percentage of glucose carbon excreted as lactate	90	80	40
Other products	Acetate Formate	Unknown	TCA cycle intermediates

4.8.5 Primate malaria

No primate malaria has been shown to possess a functional TCA cycle and the mitochondria are acristate. However, as with all *Plasmodium* species investigated, there is utilisation of oxygen involving a cyanide-sensitive cytochrome oxidase. The role of cytochrome oxidase and oxygen utilisation in a cell converting 90 per cent of the glucose substrate to lactate is as yet unresolved. There is some evidence to suggest that it is only the schizont stages that require oxygen, possibly in connection with an active TCA cycle. There is also recent evidence that the oxygen requirement is linked with pyrimidine biosynthesis (section 6.3). Primate malaria, together with all others investigated, fix CO_2, a process which is probably involved in the production of TCA cycle intermediates necessary for biosynthetic reactions. All *Plasmodium* species possess a partial phosphogluconate pathway for the production of pentoses. The energy requirement of the cell must be satisfied mainly by glycolysis, although additional ATP may be produced in the further metabolism of some intermediates. NADH is reoxidised mostly through the activity of lactate dehydrogenase, but if there is a functional electron transport chain some NADH may be reoxidised by this. Whether there is coupled respiratory chain phosphorylation is unknown.

4.8.6 Rodent malaria

Less is known about the energy metabolism of these species. Probably all mammalian stages possess acristate mitochondria with no functional TCA cycle, although the situation in mature gametocytes is uncertain. There is evidence, from inhibitor studies on the asexual blood stages, of a functional and important electron transport chain, containing cytochromes and ubiquinone components of different structures to those in the mammal and possibly consisting of two branches. A functional respiratory chain in acristate mitochondria is unusual but again is possibly linked in part with pyrimidine biosynthesis (section 6.3). The scant evidence available indicates that all stages of all *Plasmodium* species possess respiratory chains containing cytochrome oxidase and ubiquinone components different to those of the host. Analogues of ubiquinone (naphthoquinones) are effective inhibitors of electron transport and extremely toxic to malaria parasites. It is hoped that useful antimalaria drugs that do not affect the host may be developed from these compounds (section 9.2).

4.8.7 Avian malarias

These parasites possess cristate mitochondria and an active TCA cycle and undoubtedly metabolise glucose beyond pyruvate. The TCA cycle is used mainly to produce amino acids. Little CO_2 is evolved. A cytochrome system and cytochrome oxidase probably are present but of uncertain function in energy metabolism.

4.9 Other Genera

Preliminary evidence with the intraerythrocytic stages of *Babesia* suggests lactate is the main end product of glucose catabolism. Acristate mitochondria are present. Nothing is known about *Toxoplasma* and *Theileria*, although the sensitivity of the latter to menoctone, a ubiquinone analogue, strongly suggests the presence of a functional cytochrome chain.

4.10 General Conclusions

Interest in the biochemistry of parasitic protozoa is centred on the questions of how they survive in the host, and how they can be killed selectively?

Current knowledge of the energy metabolism of parasitic protozoa gives some clues as to how they cope with and take advantage of their environment. Blood-stream trypanosomes have an abundant and constant supply of both glucose and oxygen, and some species have adapted to this by evolving a seemingly simple, yet clearly very efficient, catabolism of glucose to pyruvate. This avoids the necessity of synthesising TCA cycle enzymes and respiratory chain components and hence cristate mitochondria. But why evolve the complex L-a-glycerophosphate oxidase system rather than the simple production of lactate using lactate dehydrogenase? Does lactate deleteriously affect the host? Other trypanosomes, presumably more primitive, still possess cristate mitochondria, and metabolise pyruvate further. Has the development of the intracellular stage of *Trypanosoma cruzi* in a nucleated host cell in some way been responsible for the retention of the fully oxidative metabolism? Other protozoan parasites that live in nucleated cells also possess such pathways (for example *Leishmania* and possibly *Eimeria* and *Plasmodium lophurae*). However, the mammalian malarias which live in enucleated cells provided with abundant glucose and oxygen do not.

The anaerobic protozoa, *Trichomonas* and *Entamoeba*, are presented with different problems. Reoxidation of NADH is one, the production of sufficient energy is another. NADH can be reoxidised simply by reduction of pyruvate to lactate, but these protozoa possess more complicated mechanisms. Clearly there must be reasons for this; it is attractive to suggest that there is concomitant energy production. Succinate formation is a common phenomenon in parasitic protozoa, and indeed among other anaerobic organisms. Is the reason for this common occurrence linked with energy production?

Energy metabolism provides many encouraging leads in the search for new ways of selectively killing pathogenic protozoa. The hydrogenosome and its related metabolism probably has been exploited already with the 5-nitroimidazoles, and the L-a-glycerophosphate oxidase system of trypanosomes with suramin. Surely it is possible to exploit similarly other unique systems present in *Entamoeba* and *Plasmodium*?

Our present knowledge is an interesting sample of the exciting biochemical systems that are present in parasitic protozoa. Some groups of parasitic protozoa, such as the Sporozoa, are virtually unknown entities in the biochemical sense, and even in the much studied species such as *Trichomonas vaginalis* and *Trypanosoma brucei* there is still an enormous number of unsolved problems.

4.11 Further Reading

Bowman, I. B. R. and Flynn, I. W. (1976). Oxidative metabolism of trypanosomes. In

Biology of the Kinetoplastida, vol. 1 (ed. W. H. R. Lumsden and D. A. Evans), Academic Press, New York, pp. 435–476

Muller, M. (1976). Carbohydrate and energy metabolism in *Trichomonas foetus*. In *Biochemistry of Parasites and Host–Parasite Relationships* (ed. H. van den Bossche), Elsevier-North Holland Biomedical Press, Amsterdam, pp. 3–14

Peters, W. (1969). Recent advances in the physiology and biochemistry of *Plasmodium. Trop. Dis. Bull.*, **66**, 1–29

Reeves, R. E. (1976). How useful is the energy in inorganic pyrophosphate? *Trends. biochem. Sci.*, **1**, 53–55

5 Catabolism and the Generation of Energy. II Stages outside the Vertebrate Host

5.1 Introduction

If the species is to survive, a parasite must be transmitted from one host to another. The three major mechanisms involved are: direct contact transmission (for example, *Trichomonas vaginalis*); production of resistant cysts (intestinal parasites such as *Eimeria, Toxoplasma, Entamoeba*); indirect transmission by arthropod vectors (blood parasites such as *Trypanosoma, Leishmania, Plasmodium, Babesia, Theileria*). The last two mechanisms involve environmental changes to which the parasite may have to adapt in order to survive. Thus the stages of parasitic protozoa that live outside the vertebrate host, although less significant than the vertebrate stages in terms of chemotherapy, nevertheless have been studied extensively both for their innate interest and as models of vertebrate stages. The discoveries arising from these investigations have demonstrated beautifully how some genera of parasitic protozoa adapt their metabolism to environmental conditions and that these adaptations are paralleled by morphological changes. The control of these transformations and the extent to which they occur are two of the most interesting problems currently under investigation.

5.2 Genus *Trypanosoma*

5.2.1 African trypanosomes (subgenus *Trypanozoon*)

Most research into the non-mammalian stages of trypanosomes has been on the forms that grow in axenic culture. These are morphologically and ultrastructurally similar to the insect midgut stages and are presumed, with some supporting evidence, to be biochemically equivalent. Many studies have been done on *Crithidia fasciculata*, which has biochemical and morphological similarities to the epimastigotes of *Trypanosoma brucei*. Whether the detailed metabolism of these is in fact equivalent or similar to that of any of the many vector stages, which undoubtedly differ biochemically themselves, is unknown. However, until it is possible to obtain large quantities of the vector stages, investigators must continue to use these models. The results discussed in this section are from investigations using them. Published results clearly indicate that the detailed metabolism varies with the culture conditions. In this section an attempt has been made to give an overall summary.

In contrast to the blood-stream stages, culture forms of *T. brucei* possess mitochondria with numerous plate-like cristae. Oxygen utilisation is cyanide sensitive in part. Substrate utilisation also differs. Although glucose normally will support growth quite adequately, and is the preferred substrate with some species, it has been demonstrated that with some strains of *T. brucei* in culture the amino acid proline is the favoured substrate. This correlates well with the change in environment, for the contents of the tsetse fly midgut are about 20 per cent protein but only 0.1 per cent carbohydrate. Tsetse flies also contain very high concentrations of proline in their haemolymph and this is the major energy source of their flight muscles. This preference for proline is a recent discovery, and much of our knowledge of the energy metabolism of trypanosome culture forms has come from studies of the protozoa growing on glucose. Other substrates have been shown also to support growth of culture form trypanosomatids, especially hexoses and glycerol. *Crithidia fasciculata* will grow also on both pentoses and various alcohols. The transport of substrates into the culture forms is similar to the transport into the blood-stream stages, though there is recent preliminary evidence to suggest that in some strains proline is taken up more readily than glucose.

Proline catabolism

Proline is catabolised, mainly to CO_2, alanine and aspartate, by the pathway shown schematically in figure 5.1. When *T. brucei* culture forms are growing on proline they apparently lack a fully functional TCA cycle. However, the partial TCA cycle present accounts for the CO_2 evolved and the energy produced, for the NADH generated is presumed to be reoxidised by the respiratory chain with the concomitant synthesis of ATP. The enzyme alanine aminotransferase is present at a very high concentration and is critical to growth on proline. Its presence explains the appearance of alanine among the end products of metabolism. Aspartate is produced from oxaloacetate in a second transamination reaction, catalysed by aspartate aminotransferase, producing α-ketoglutarate which can re-enter the partial TCA cycle. Pyruvate carboxylase and the malic enzyme, not yet investigated in *T. brucei* culture forms growing on proline, are known to be active in the same organism growing on glucose. The mechanism of the conversion of proline to glutamate has not been investigated in *T. brucei*. In mammals it occurs as indicated in figure 5.1. Other amino acids are utilised by *T. brucei* in culture, although none as the main carbon and energy source. Threonine is consumed in large quantities and broken down to glycine and acetate units. The latter are the major source of two-carbon units for fatty acid synthesis (section 8.4). Some bacteria apparently can grow on threonine as the sole energy, carbon and nitrogen source. Such an ability has not been found among the parasitic protozoa.

Glucose catabolism

Most of the studies of the metabolism of culture form trypanosomatids have involved organisms grown on glucose as the major energy and carbon source.

The main end products of this catabolism are CO_2, succinate, acetate, alanine and pyruvate, the amount of each varying with the species and strain. Pyruvate is produced by the breakdown of glucose through the glycolytic pathway. The regulation of this pathway in Crithidia is unusual and does not involve phosphofructokinase. The importance of the phosphogluconate pathway is un-

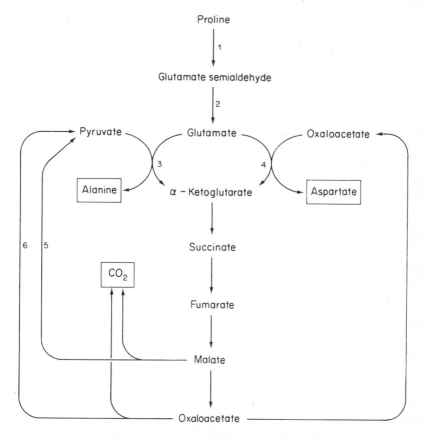

Figure 5.1 Catabolism of proline in *Trypanosoma brucei* culture forms. End products are in boxes. Enzymes: 1, L-proline oxidase; 2, glutamate semialdehyde dehydrogenase; 3, alanine aminotransferase; 4, aspartate aminotransferase; 5, malic enzyme; 6, pyruvate carboxylase.

clear, but it probably exists for the synthesis of pentoses and NADPH. In aerobic conditions, pyruvate can enter the TCA cycle and be fully oxidised to CO_2. The NADH concomitantly produced is reoxidised by a respiratory chain. However, even in aerobic conditions, and certainly if anaerboic conditions occur, both PEP and pyruvate are also carboxylated to oxaloacetate and malate, which, by the action of the TCA cycle enzymes acting in reverse, are reduced to succinate, which is excreted. The enzymes catalysing the carboxylation reactions are pyruvate carboxylase, PEP carboxykinase (GTP) and the malic enzyme (section 4.5).

Interestingly the decarboxylation activity of the malic enzyme, which would convert malate to pyruvate, is inhibited by acetyl CoA and oxaloacetate, at close to physiological concentrations. This ensures the production of succinate in conditions of high energy availability and provides a second mechanism whereby excess NADH can be reoxidised. This is necessary because it appears that the activity of the electron transport chain is not sufficient to cope with the flux through the glycolytic pathway. The possibility exists, also, although there is no evidence as yet, that the conversion of fumarate to succinate, catalysed by a particulate enzyme fumarate reductase, may be linked to energy production (section 4.5). The excretion of alanine by the culture form trypanosomatids growing on glucose indicates that some pyruvate must undergo transamination, a reaction catalysed by alanine aminotransferase. Whether there is an advantage in excreting alanine when growing on glucose is unclear. A possible explanation is that this production is a consequence of the presence of large quantities of alanine aminotransferase, an enzyme which is required and synthesised mainly for growth on proline. Does this indicate evolution from an organism growing on proline? Some pyruvate is converted to acetate by an unknown mechanism. However, the enzyme carnitine acyltransferase has been found in the blood-stream forms of *T. brucei*, which suggests that a related pathway may be reponsible for the production of acetate in culture form *T. brucei*

$$\text{acetyl CoA} + \text{carnitine} \longrightarrow \text{acetyl carnitine} + \text{CoA}$$

$$\text{acetyl carnitine} \longrightarrow \text{acetate} + \text{carnitine}$$

Our understanding of the catabolism of glucose is outlined schematically in figure 5.2.

The utilisation of oxygen by culture form trypanosomatids is cyanide sensitive to some extent. The degree of sensitivity varies with the species, strain and culture conditions, but in the majority of cases is less than 100 per cent. The cyanide sensitive respiration involves an electron transport chain containing cytochromes b, c and $a + a_3$. Although similar to the cytochromes present in mammals, they differ in detail. The cytochrome c of culture form trypanosomatids has been investigated in depth. It differs from mammalian cytochrome c in the wavelength of the α absorption maximum of the reduced form (555–558 nm as opposed to 550 nm for mammalian cytochrome c) and also in the mechanism of binding of the haem prosthetic group. The primary structures differ at several positions. It is reported that cytochrome c of *C. fasciculata* differs so markedly in primary structure from that of *C. oncopelti* that the two species are phylogenetically quite distinct. The significance of this finding is uncertain.

The cyanide-insensitive oxygen utilisation has been the subject of much research and controversy. It is known to be localised in the mitochondria, but the L-α-glycerophosphate oxidase system that is found in the blood-stream trypanosomes is present only at low levels in culture trypanosomes and can be responsible for only a proportion of the cyanide-insensitive oxygen utilisation. The present evidence indicates that a cytochrome o, previously identified only in prokaryotic organisms, acts as the terminal acceptor of an electron transport chain and is responsible for the remainder of the cyanide-insensitive respiration. The present knowledge of the compositions of the two electron transport chains apparently present in culture form trypanosomes are shown in figure 5.3. The ubiquinone of *C. fasciculata* differs from that in the mammal, and so

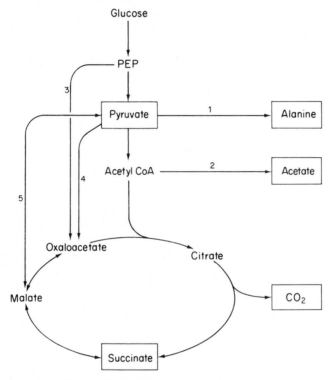

Figure 5.2 Catabolism of glucose in *Trypanosoma brucei* culture forms. End products are in boxes. Enzymes: 1, alanine aminotransferase; 2, carnitine acyltransferase; 3, PEP carboxykinase; 4, pyruvate carboxylase; 5, malic enzyme.

is potentially a target for chemotherapeutic attack. It is interesting, but unexplained, that neither rotenone nor amytal affect trypanosomatid electron transport chains. Whether the two terminal electron acceptors are parts of distinct chains or simply different branches of the same chain (as shown in figure 5.3) is unknown. Intact culture cells have an apparent K_m of 0.1 μM for O_2, which is much lower than that of blood-stream *T. brucei*, which have a K_m of greater than 2 μM for O_2. This helps to explain the necessity of the cytochrome chains for the survival of the organism in the insect vector, because the oxygen tension in the gut of the tsetse fly is much lower than that in mammalian blood.

Respiratory chain phosphorylation is coupled to electron transport, but there seem to be only two phosphorylation sites, equivalent to sites II and III of the mammalian system. The mitochondrial ATPase of *Crithidia* has been characterised and is markedly different to the equivalent mammalian enzyme in its susceptibility to inhibitors. This is a possible target for chemotherapeutic attack, especially as it is one of the few mitochondrial enzymes that persists in the mitochondria of some strains of blood-stream trypanosomes. However, it is doubtful whether it is vital to these stages. Thus both energy production and NADH reoxidation are achieved by this coupled respiratory chain phosphorylation and electron transport system.

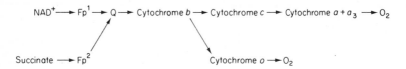

Figure 5.3 Schematic representation of the proposed respiratory chain in *Trypanosoma brucei* culture forms.

Transformation of blood-stream forms to culture forms

The metabolism of the short stumpy blood-stream trypomastigotes has been found to be intermediate between the metabolism of the long slender blood-stream form and the culture form (table 5.1). The end products are acetate, glycerol, pyruvate, succinate and CO_2, the mitochondria possess tubular cristae, and in many ways the metabolism appears similar to that of blood-stream *T. vivax* and *T. congolense* (section 4.3). These short stumpy

Table 5.1 Comparison of the metabolic status of long slender and short stumpy blood-stream and culture forms of *Trypanosoma* species of subgenus *Trypanozoon*

	Blood-stream forms		Culture form
	Long slender	Short stumpy	
Glycolysis	+	+	+
TCA cycle	—*	Partial	+†
L-α-glycerophosphate oxidase	+	+	Low
Proline oxidase	—	Low	+
Cytochrome chain	—	—	+
Cyanide sensitivity	—	—	Partial
Mitochondria	Very few tubular cristae	Tubular cristae	Plate-like cristae
Growth temperature (°C)	37	37	25

+, present; —, absent
*Recent work suggests some of the enzymes may be present in some strains (but not the key citrate synthase and succinate dehydrogenase).
†The cycle may be only partial in some strains when growing on proline (see text).

trypomastigotes occur mostly when the infection is in remission (when the host is recovering from an acute attack of trypanosomiasis and the parasite numbers are declining) and they are thought to be the only stages of the blood-stream parasite that are infective to the tsetse fly, and so are responsible for transmission of the infection. The demonstration that the metabolism of this form is intermediate between the long slender trypomastigote and the culture form shows that transformation begins in the blood stream. However, it is completed only if these short stumpy forms are taken up by a vector in a blood meal (or cultured *in vitro*). The stimulus that causes transformation to begin is unknown, but clearly the immune response of the host may be responsible in part.

5.2.2 *Trypanosoma vivax* and *T. congolense*

Little is known of the vector stages of these parasites. However, they do possess mitochondria with plate-like cristae indicating that similar, if less dramatic, changes as in *T. brucei* might occur during their life cycles.

5.2.3 *Trypanosoma cruzi*

The energy catabolism of the culture epimastigote forms of *T. cruzi* (which are most likely equivalent to the main developmental stage in the gut of the vector) is similar to that of the blood-stream trypomastigotes and intracellular amastigotes. These culture forms possess mitochondria with a few plate-like cristae. Glucose can be utilised but it is unclear whether this is the major or even preferred energy source; amino acids such as proline also can be used. The glycolytic pathway is fully active, the TCA cycle is functional. Respiration is cyanide sensitive in part, involves a poorly characterised cytochrome chain, and is coupled to respiratory chain phosphorylation. The mitochondrial ATPase is broadly similar to that present in mammals. The major end products of glucose catabolism are CO_2 (about 32 per cent), acetate (about 20 per cent) and succinate (about 34 per cent); the mechanisms by which these are produced are probably similar to those of the blood-stream forms (subsection 4.3.7). Threonine is consumed probably as a source of acetate units for fatty acid synthesis similarly to *T. brucei* culture forms. Clearly the metabolic switch associated with the transformation of *T. cruzi* blood-stream forms to culture forms is less dramatic than that occurring in *T. brucei*, indicating that *T. cruzi* may be the more primitive organism.

5.3 Genus *Leishmania*

In the sand fly vector, *Leishmania* grows as the promastigote form. *Leishmania* also can be grown in culture *in vitro* in a morphologically similar form and these culture promastigotes have been used in investigations of the energy catabolism. They possess many large mitochondria with plate-like cristae, a functional TCA cycle and an active cytochrome chain. Oxygen utilisation is sensitive to antimycin A and cyanide, but whether this sensitivity is total is unclear. Both glucose and proline support growth of promastigotes, although with some species the latter is the preferred substrate, glucose only being metabolised when the culture reaches the stationary phase. Proline is catabolised by very similar pathways to those used by *Trypanosoma brucei* culture forms (section 5.2), alanine being the major organic end product. Glucose, which is taken up by a mediated transport mechanism, is catabolised through the glycolytic and phosphogluconate pathways. The major end products are CO_2 and succinate; small amounts of pyruvate also are excreted. Succinate production involves carboxylation of PEP and pyruvate to oxaloacetate and malate, and the reduction of these intermediates by the TCA cycle enzymes acting in reverse. The energy requirements of the organism are presumably mainly met by respiratory chain phosphorylation coupled to the electron transport chain. NADH oxidation is achieved by the same system and by the production of succinate. The energy metabolism of *Leishmania* promastigotes and *T. brucei* culture forms are similar in many ways. This is probably a reflection of the similarity of their environments. Whether similarities exist also in the way in which their metabolisms vary between different stages of their life cycles will become clearer when our knowledge of the metabolism of the *Leishmania* amastigotes increases (section 4.4). Preliminary evidence suggests that there is less variation between the different developmental stages of *Leishmania*.

5.4 Genus *Entamoeba*

The biochemical changes which occur during the encystment of the trophozoite stages of *Entamoeba histolytica* are linked with the formation of the cyst wall. The biochemistry of the resting cysts, however, has not been investigated in detail, although they are known to possess large glycogen stores and to respire at very low rates.

5.5 Genus *Eimeria*

Both the oocyst and the sporozoite contain stores of polysaccharide, similar to amylopectin, and lipids. Amylopectin phosphorylase and β-galactosidase have been demonstrated in the oocyst. Both stages possess cristate mitochondria and respire vigorously; oxygen utilisation is cyanide sensitive to an extent and is at least partially mediated by a cytochrome system. This system in the oocyst apparently contains cytochromes *b* and *o* (cytochrome *c* has not been observed) and is sensitive to antimycin A and cyanide. The detailed composition of the chain is yet to be elucidated. Respiration by mitochondria isolated from oocysts is inhibited by anticoccidial quinolones, which seem to act on the cytochrome chain around ubiquinone and cytochrome *b*. The observed lesser sensitivity of mitochondria from strains of *Eimeria* resistant to the drugs and the lack of sensitivity of those from avian cells, confirm that the primary site of action of the drugs is the inhibition of electron transport (section 9.2). Both the oocyst and the sporozoite normally catabolise glucose aerobically; CO_2 is the major end product. However, both can survive anaerobically by catabolising their reserves of polysaccharide mainly to lactate, although a little glycerol and CO_2 are produced also. The production of CO_2 may be from the phosphogluconate pathway but could be explained by the presence of small amounts of oxygen in the environment. Sporulation of the oocyst does not occur in the absence of oxygen. Lack of information about trophozoite and schizont stages (section 4.7) does not allow any firm conclusions to be made concerning the variation of energy metabolism during the life cycle, although preliminary histochemical evidence suggests a situation similar to that in mammalian malaria parasites (section 5.6).

5.6 Genus *Plasmodium*

Very little is known of the metabolism of the stages of *Plasmodium* that grow in the mosquito. Ultrastructural and histological investigations of *P. berghei* have shown that mature gametocytes, the oocyst and sporozoite possess mitochondria with cristae and some TCA cycle enzymes signifying that there probably is an active TCA cycle and electron transport chain. This contrasts with the intraerythrocytic stages of the same species which contain acristate mitochondria (section 4.5) and suggests, as in the *T. brucei* group, the presence of biochemical changes associated with development in the vector. Avian malarias are rather different, in that all stages possess active cristate mitochondria.

5.7 Other Genera

Nothing is known of the energy metabolism of stages of *Toxoplasma*, *Babesia* and *Theileria* that develop outside the vertebrate host.

5.8 General Conclusions

It is clear from this account that there is an incomplete picture of the biochemistry of the different stages which occur during the life cycles of most parasitic protozoa. These gaps will remain until techniques have been developed either to isolate all stages of the parasites in quantities sufficient for biochemical study, or to culture the parasites axenically *in vitro*. Nevertheless, it is clear that important morphological and biochemical changes can occur, especially in parasites whose transmission is indirect, involving arthropod vectors. In some cases, these are major, as with the African trypanosomes and mammalian malaria parasites, whereas in others they are less dramatic, such as those that probably occur in *Trypanosoma cruzi*, *Leishmania* and avian malaria parasites. The different responses of organisms to an altered environment are probably a reflection of the evolutionary stage reached by the parasite in its biochemical adaptation to its host. The mechanisms by which these morphogenetic and biochemical transformations are triggered and controlled are unknown, but comprise one of the most interesting areas of modern research in protozoology.

5.9 Further Reading

Bowman, I. B. R. (1974). Intermediary metabolism of pathogenic flagellates. In *Trypanosomiasis and Leishmaniasis with Special Reference to Chagas' Disease*, Ciba Foundation Symposium No. 20 (new series), Associated Scientific Publishers, Amsterdam, pp. 255–284

Evans, D. A. and Brown, R. C. (1972). The utilisation of glucose and proline by culture forms of *Trypanosoma brucei. J. Protozool.*, **19**, 686–690

Hill, G. C. (1976). Characterisation of electron transport systems present during the life cycle of African trypanosomes. In *Biochemistry of Parasites and Host–Parasite Relationships* (ed. H. Van den Bossche), Elsevier-North Holland Biomedical Press, Amsterdam, pp. 31–50

Howells, R. E. and Maxwell, L. (1973). Further studies on the mitochondrial changes during the life cycle of *Plasmodium berghei*: electrophoretic studies on isocitrate dehydrogenase. *Ann. trop. Med. Parasit.*, **67**, 279–283

Krassner, S. M. (1969). Proline metabolism in *Leishmania tarentolae. Expl Parasit.*, **24**, 348–363

Trigg, P. I. and Gutteridge, W. E. (1977). Morphological, biochemical and physiological changes occurring during the life cycles of protozoan parasites. In *Parasite Invasion*, Symposium of the British Society for Parasitology, vol. 15 (ed. A. E. R. Taylor and R. Muller), Blackwell Scientific Publications, Oxford, pp. 57–81)

6 Nucleic Acid Metabolism

6.1 Introduction

Nucleic acids are polymers of nucleotides. The latter are composed of a five-carbon sugar (D-ribose or D-deoxyribose) linked both to a nitrogen base (purine or pyrimidine) and a phosphate group (nucleosides are nucleotides without this phosphate group). The purine and pyrimidine nucleotides which occur most frequently in the nucleic acids of cells are illustrated in figure 6.1. In the unpolymerised state, they occur as nucleoside diphosphates (NDP) and triphosphates (NTP) as well as the monophosphate forms illustrated. In the polymerised state, the phosphate on the 5'-OH of the sugar is linked to the 3'-OH of the adjacent sugar residue so that nucleic acids consist of a backbone of alternating residues of sugar and phosphate with the bases projecting out from this. Two types of nucleic acid exist, deoxyribonucleic acid (DNA) which is a polymer of deoxyribonucleotides and ribonucleic acid (RNA) which is a polymer of ribonucleotides. Nucleic acid metabolism is concerned with the synthesis and breakdown of purine and pyrimidine nucleotides, their utilisation in DNA and RNA synthesis and the structure, localisation and function of these nucleic acids.

6.2 Purine Metabolism

In mammals there are two sources of purine nucleotides. The liver synthesises the purine ring *de novo* from glycine, formate, CO_2, glutamine and aspartate as detailed in figure 6.2. Some other body organs appear to salvage purine bases and nucleosides, which have been released into the blood stream by the liver, according to the scheme detailed in figure 6.3. Whatever their origin, once assimilated, the bases and nucleosides are usually freely interconvertible (figure 6.3).

All the parasitic protozoa examined to date, with two possible exceptions (*Trypanosoma cruzi* and *Crithidia oncopelti*), appear to be unable to synthesise the purine ring *de novo* and are thus dependent on the host for a source of preformed purine. It must be noted, however, that so far only *Trypanosoma* and *Plasmodium* have been examined in any detail in this respect; information concerning *Leishmania, Trichomonas, Entamoeba, Eimeria, Toxoplasma* and *Theileria* is scanty and there is no information for *Babesia*. Even with

DEOXYRIBONUCLEOTIDES | RIBONUCLEOTIDES

Purines

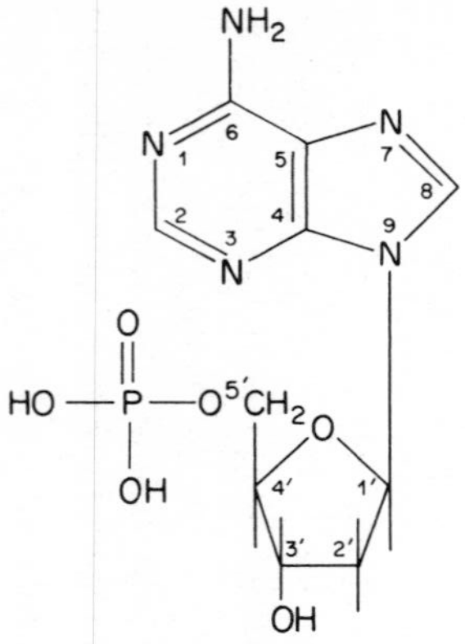

Deoxyadenosine monophosphate (dAMP)

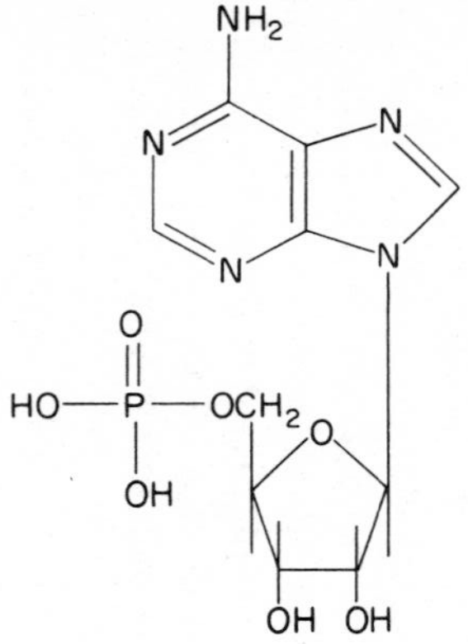

Adenosine monophosphate (AMP)

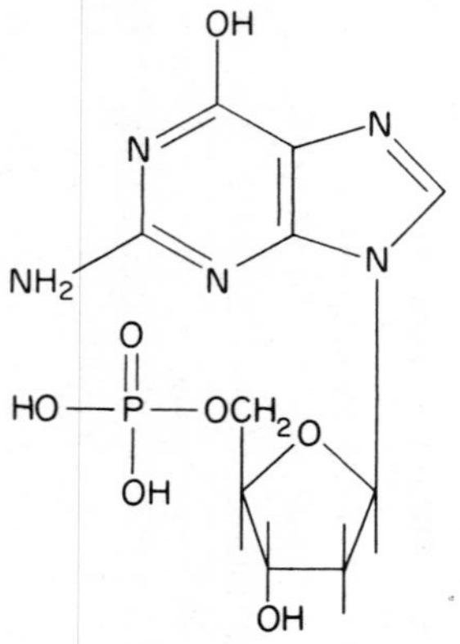

Deoxyguanosine monophosphate (dGMP)

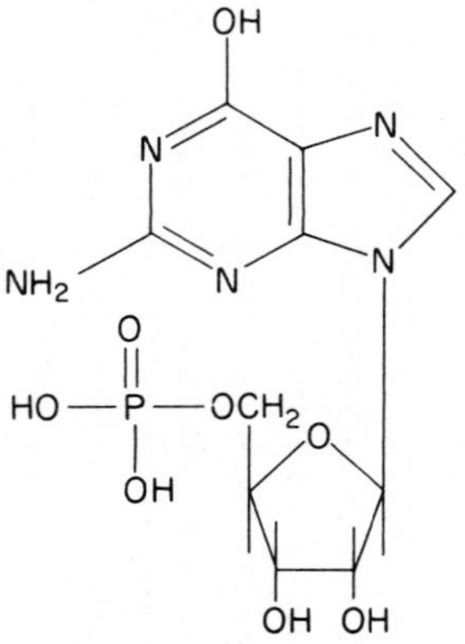

Guanosine monophosphate (GMP)

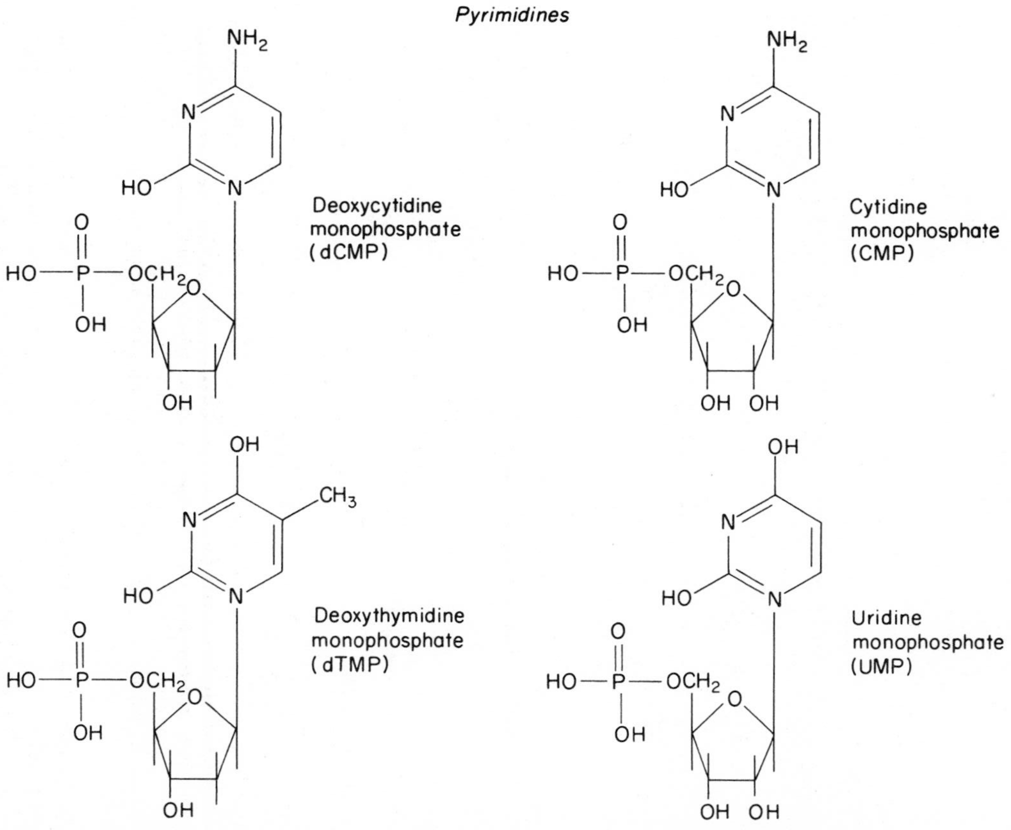

Figure 6.1 Chemistry of nucleotides.

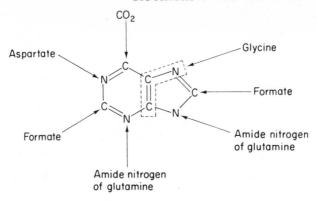

Figure 6.2 Origins of the atoms of the purine ring.

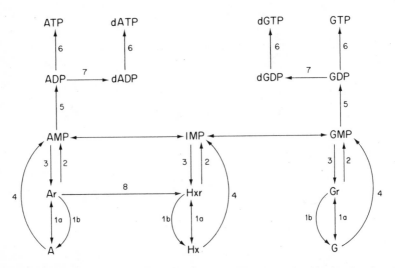

Figure 6.3 Salvage and interconversion of purines. Enzymes: 1a, nucleoside phosphorylase; 1b, nucleoside hydrolase; 2, nucleoside kinase; 3, nucleotide phosphatase; 4, phosphoribosyltransferase; 5, NMP kinase; 6, NDP kinase; 7, ribonucleotide reductase; 8, adenosine deaminase. IMP, inosine monophosphate; Hxr, inosine; Hx, hypoxanthine; Ar, adenosine; A, adenine; Gr, guanosine; G, guanine; other abbreviations as in figure 6.1.

Trypanosoma and *Plasmodium*, evidence for the conclusion has come mainly from studies on the composition of minimal defined media that will support growth of parasites and from a failure of radiotracers such as [14C]-glycine and [14C]-formate to label the purine bases of the nucleic acids of the parasites. In no instance has a detailed study been undertaken to establish a lack of the necessary enzymes.

One possible exception to the general rule that parasitic protozoa cannot synthesise the purine ring *de novo* is *Trypanosoma cruzi* for which there is some evidence for the incorporation of [14C]-glycine into purines by intracellular stages growing in tissue cells. This remains to be substantiated. The other possible exception concerns *Crithidia oncopelti*, a model trypanosomatid flagellate (section 3.2). This organism is grown in a defined medium containing purine (see section B4) but it will readily incorporate [^{14}C]-glycine into nucleic acid purine so that there is no doubt here that it can synthesise purines *de novo*. However, unusual organelles have been described in its cytoplasm (see appendix C) and it seems likely that these 'bipolar bodies' in fact represent endosymbiotic bacteria which synthesise nutrients for the host. The ability to synthesise purine *de novo* may, therefore, not be a property of the protozoan itself.

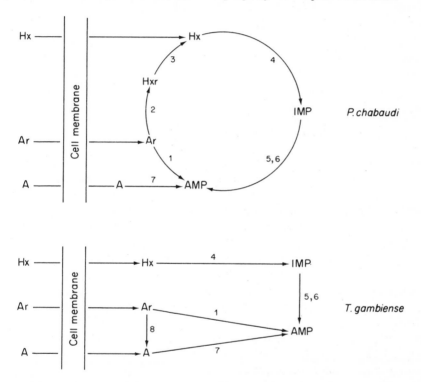

Figure 6.4 Purine salvage in *Plasmodium chabaudi* and *Trypanosoma gambiense*. Enzymes: 1, adenosine kinase; 2, adenosine deaminase; 3, inosine phosphorylase; 4, hypoxanthine phosphoribosyltransferase; 5, adenylosuccinate synthetase; 6, adenylosuccinate lyase; 7, adenine phosphoribosyltransferase; 8, purine nucleoside hydrolase. Abbreviations as for figures 6.1 and 6.3.

Absence of the ability to synthesise purines *de novo* implies dependence on a salvage pathway. Studies with minimal defined media on parasites in culture and on the incorporation of radioactive purines have provided good evidence for the existence of purine salvage in *Trypanosoma* and *Plasmodium* and scanty evidence for this in *Leishmania*, *Trichomonas*, *Entamoeba*, *Eimeria*, *Toxoplasma* and *Theileria*. There is no information for *Babesia*. Where

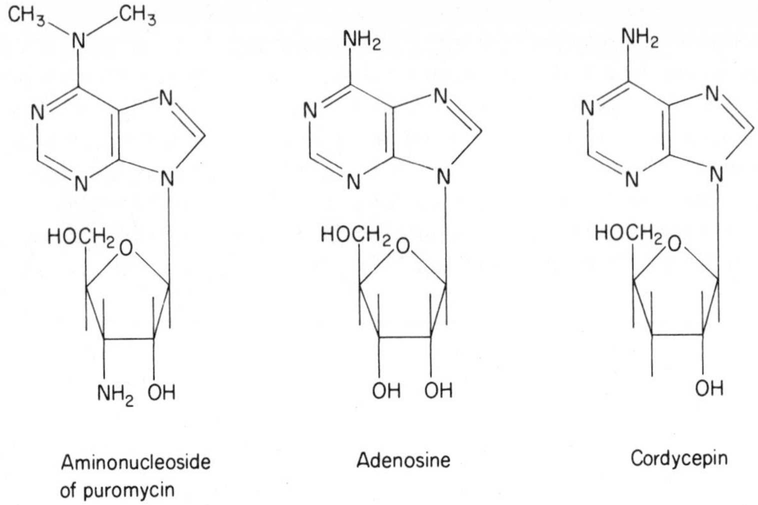

Figure 6.5 Chemical structures of purine analogues.

salvage does occur, it appears to be by pathways similar to those outlined in figure 6.3 and the different purines seem to be freely interconvertible. Only in *Plasmodium* and *Trypanosoma gambiense* have extensive enzymic and metabolic studies been done to determine the details of purine salvage.

In *Plasmodium*, at least three pathways exist, involving, in order of preference, the salvage of hypoxanthine, adenosine and adenine (figure 6.4). Only a very small amount of adenine salvage occurs. Note that adenosine can either be converted to AMP by adenosine kinase or to inosine by adenosine deaminase and thence to hypoxanthine by an inosine phosphorylase. The preferred metabolic route appears to be via deamination and phosphorolysis. These initial reactions probably occur outside of or on the membrane of the parasite; it is likely that it is hypoxanthine which crosses the parasite membrane and is then further metabolised to AMP. Note that enzymes involved in purine metabolism are also present in the cytoplasm of the host red cell.

Purine salvage in *T. gambiense* involves three pathways and is similar in many ways to that in *Plasmodium* (figure 6.4). The main difference involves adenosine salvage; in *T. gambiense* adenosine can be either converted to AMP by adenosine kinase (as in *Plasmodium*) or degraded to adenine by a purine nucleoside hydrolase (instead of being deaminated to inosine). There is also some recent evidence that *T. brucei* and *T. rhodesiense* can salvage purine nucleotides, especially AMP, without prior dephosphorylation. This is a rare ability; most cells are impermeable to nucleotides.

One pathway involving the metabolism of ATP not shown in figure 6.3 is the formation of adenosine-3',5'-monophosphate (cyclic AMP). The occurrence and regulatory role of this nucleotide are well documented for many microorganisms and animal tissues. The intracellular level of cyclic AMP is a function of the relative activities of adenylate cyclase which synthesises it and $3',5'$-cyclic AMP phosphodiesterase which degrades it. Both enzymes have been detected in *T. gambiense*, along with cyclic AMP itself. The regulatory role of this nucleotide is discussed in subsection 7.5.7.

Dependence on the salvage of purines might be expected to render parasitic protozoa sensitive to the cytotoxic action of purine analogues. Many of these have been developed in recent years as possible antitumour drugs. Two of those which have been shown to have activity against at least one genus of parasitic protozoan, either *in vivo* or *in vitro*, are detailed in figure 6.5. None tried so far has sufficient activity to be useful in veterinary or clinical infections.

6.3 Pyrimidine Metabolism

In mammals, pyrimidines are synthesised *de novo* from glutamine, bicarbonate and aspartate (figure 6.6). The product of this pathway is UMP which is then converted to other pyrimidines by the reactions detailed in figure 6.7. Some tissues and cells can also salvage pyrimidines (figure 6.7).

All parasitic protozoa examined so far (*Trypanosoma*, *Crithidia*, *Leishmania*, *Entamoeba*, *Eimeria*, *Toxoplasma* and *Plasmodium*) appear to be able to synthesise *de novo* at least part of their pyrimidine requirements. The evidence for this conclusion has come mainly from studies of the composition of minimal defined media for the growth of organisms and the incorporation of radioactive bicarbonate and orotate into nucleic acid pyrimidine rings. In addition, some of the individual enzymes of the pathway have been described in some species, but only in *C. fasciculata* and *Plasmodium* have all the enzymes of the pathway been shown to be present. No information is available as yet about pyrimidine metabolism in *Trichomonas*, *Babesia* and *Theileria*.

It has been claimed that *Trypanosoma cruzi* is an exception to the general rule that all parasitic protozoa appear to be able to synthesise the pyrimidine ring *de novo*. It has been suggested that all forms of *T. cruzi* rely on salvage. There is now evidence to suggest that this conclusion is wrong, at least for culture epimastigote and blood trypomastigote forms; they appear to be able both to synthesise and to salvage pyrimidines and are thus not an exception to the general rule.

Of particular interest in the *de novo* pyrimidine biosynthetic pathway are the enzymes which convert dihydroorotate to orotate. The mammalian enzyme is a dehydrogenase and is mitochondrial, particulate, irreversible and intimately connected to the respiratory cytochrome chain to which it passes electrons directly, probably at the ubiquinone level. There is no involvement of pyrimidine nucleotides or unconjugated pteridine cofactors (figure 6.8). It has been shown recently that in *Plasmodium*, formation of orotate occurs by a similar mechanism, thus providing an explanation for the existence of an electron transport chain in species such as *P. knowlesi* in which glucose is not generally metabolised beyond lactate (section 4.8). Malaria parasites seem to contain ubiquinones different from those synthesised in mammals (section 9.2) and thus this reaction might be a target for selective drug therapy. In contrast to this

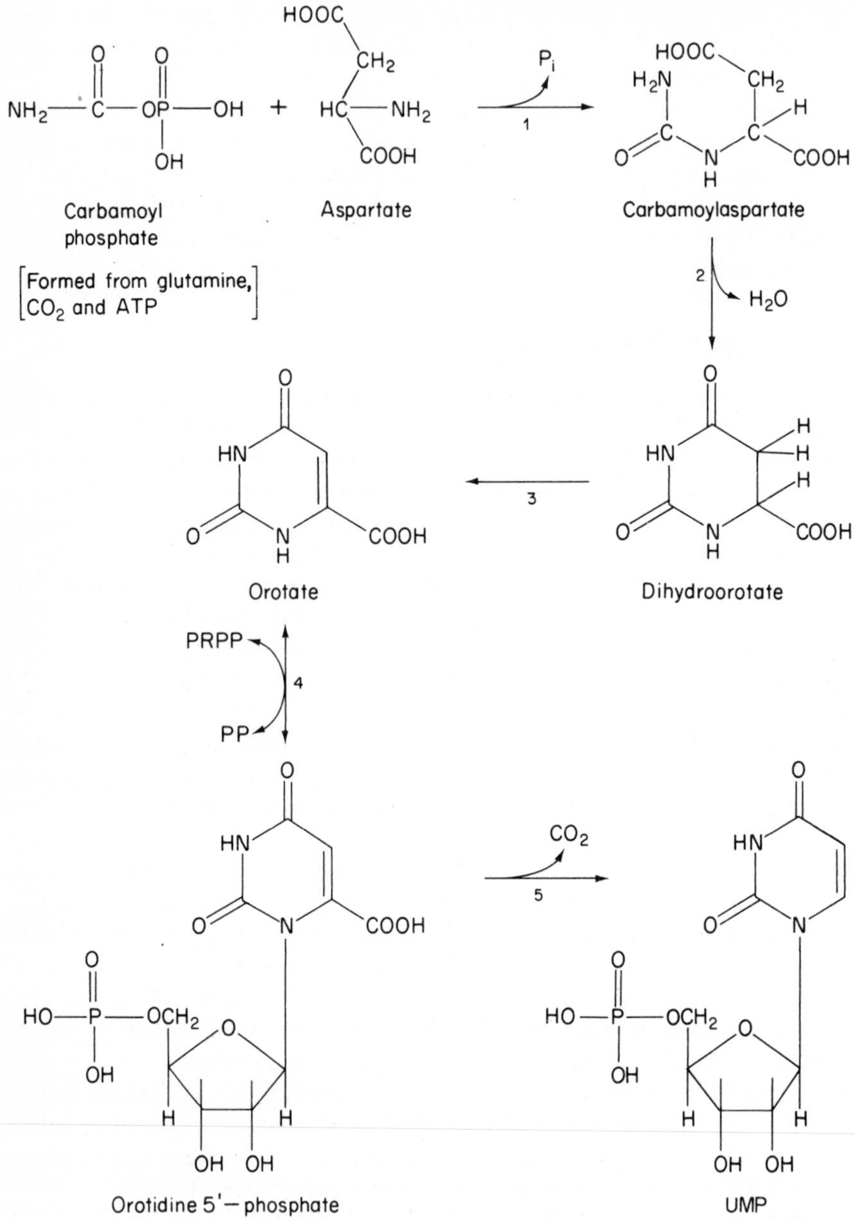

Figure 6.6 Pyrimidine biosynthesis *de novo* in mammals. Enzymes: 1, aspartate carbamyltransferase; 2, dihydroorotase; 3, dihydroorotate dehydrogenase; 4, orotidine 5'-phosphate pyrophosphorylase; 5, orotidine 5'-phosphate decarboxylase. PRPP, phosphoribosylpyrophosphate.

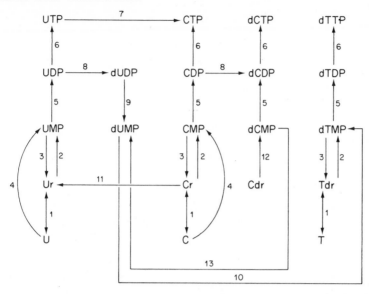

Figure 6.7 Salvage and interconversion of pyrimidines. Enzymes: 1, nucleoside phosphorylase; 2, nucleoside kinase; 3, NMP phosphatase; 4, phosphoribosyltransferase; 5, NMP kinase; 6, NDP kinase; 7, CTP synthetase; 8, ribonucleotide reductase; 9, NDP phosphatase; 10, thymidylate synthase; 11, cytidine deaminase; 12, deoxycitidine kinase; 13, deoxycytidylate aminohydrolase. U, uracil; Ur, uridine; C, cytosine; Cr, cytidine, Cdr, deoxycytidine T, thymine; Tdr, thymidine; other abbreviations as in figure 6.1.

situation in *Plasmodium* (and possibly other sporozoa such as *Eimeria* and *Toxoplasma*) and mammalian cells,* the enzyme in the Kinetoplastida (*Trypanosoma*, *Crithidia* and *Leishmania*) is a hydroxylase and is soluble, irreversible, reacts with oxygen directly and requires tetrahydrobiopterin as cofactor. NAD is not involved directly but is required in the reduced form to recycle the pteridine cofactor (figure 6.8). The hydroxylase is inhibited by orotate, but not by pyrimidine or purine nucleotides; orotate thus contols its own biosynthesis. The mechanism of the reaction in *Entamoeba* has yet to be established.

In addition to a biosynthetic pathway *de novo*, a salvage pathway for pyrimidines appears to be present in most parasitic protozoa, though its importance varies with the genus. In *Trypanosoma*, it appears to rank about equal in importance with the *de novo* pathway. In *Plasmodium*, although the host red cell is freely permeable to pyrimidines, salvage is of minimal importance so that it is impossible to label to any extent malaria nucleic acids with uracil, uridine, cytosine or cytidine. Pyrimidine salvage appears to occur in *Trichomonas En-tamoeba*, *Eimeria* and *Toxoplasma*, but insufficient study has been made for its importance to be established. No information is available about it in *Leishmania*, *Babesia* and *Theileria*. The overall pathways seem to be similar to those detailed in figure 6.7, though the nucleosides seem to be more freely utilised than the free bases. Studies on the fate of radioactive pyrimidines after

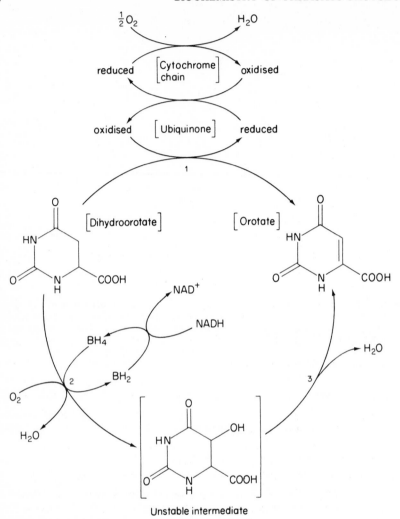

Figure 6.8 Conversion of dihydroorotate to orotate in mammals and *Plasmodium* (upper route) and *Trypanosoma*, *Crithidia* and *Leishmania* (lower route). Enzymes: 1, dihydroorotate dehydrogenase; 2, dihydroorotate hydroxylase; 3, spontaneous. BH_4, tetrahydrobiopterin; BH_2, dihydrobiopterin.

salvage suggest that they are freely interconvertible, presumably in the way illustrated in figure 6.7. Pyrimidine analogues such as 5-fluorouracil show antiprotozoan activity *in vitro* and sometimes *in vivo*, but all are too toxic for clinical or veterinary use.

6.4 Deoxyribonucleotide Metabolism

In mammalian cells and most bacteria, deoxyribonucleotides required for

DNA synthesis are made from the corresponding ribonucleoside diphosphates by the enzyme complex ribonucleotide reductase. This complex consists of four separate enzyme proteins, one of which, thioredoxin, a heat-stable flavoprotein, is the immediate electron donor in the reaction. In some bacteria, however, such as the *Lactobacilli*, the reduction is made at the nucleoside triphosphate level by an enzyme complex which probably includes thioredoxin and which requires deoxyadenosylcobalamin as cofactor.

Deoxyribonucleotide formation in parasitic protozoa has been little studied. It has been inferred, however, that in *Crithidia* reduction is carried out at the ribonucleoside diphosphate level since cultures can be synchronised with hydroxyurea, which is believed to be a specific inhibitor of this type of reaction. It is also likely that in *Trypanosoma gambiense* at least, deoxycytidine, which is known to be a major deoxyribonucleoside in circulating blood, is taken up and converted to deoxycytidine monophosphate (dCMP) by a deoxycytidine kinase (figure 6.7). There is also some evidence of a small amount of salvage of a similar nature in *P. chabaudi*, the dCMP thus formed in this species possibly being further metabolised to deoxyuridine monophosphate (dUMP) by deamination (figure 6.7, reaction 13).

Uridine diphosphate (UDP) is converted by ribonucleotide reductase to dUDP, rather than to a thymine nucleotide which is required for DNA synthesis. The necessary methylation of the pyrimidine ring is carried out in most cells at the nucleoside monophosphate level and involves the conversion of dUMP to dTMP (deoxythymidine monophosphate) by the enzyme thymidylate synthase. This enzyme requires N^5,N^{10}-methylene tetrahydrofolate as cofactor which, in the course of the reaction, is converted to dihydrofolate. It is reformed by the enzyme dihydrofolate reductase and recharged with an active one-carbon transfer group by one of a number of possible methyltransferase enzymes (figure 6.9).

It is likely, although only proven so far for *Trypanosoma*, *Eimeria* and *Plasmodium*, that a similar cycle exists in all parasitic protozoa. Thymidylate synthase has been isolated from both *Trypanosoma* and *Plasmodium* and the enzyme in *Trypanosoma* has been studied in detail recently. It differs from the isofunctional mammalian enzyme in being inhibited by Mg^{2+} and having a greater sensitivity to inactivation by mercaptoethanol, higher apparent K_m for both substrate and cofactor, a higher apparent molecular weight (200 000 daltons compared to 60 000 daltons) and a greater sensitivity to inhibition by the African trypanocide, suramin (section 9.2). Dihydrofolate reductase has been isolated from *Trypanosoma*, *Eimeria* and *Plasmodium*. It too differs from the isofunctional vertebrate enzyme, having different cofactor requirements (NADH can be used as well as NADPH), higher apparent molecular weight (about 200 000 daltons compared to 20 000–30 000 daltons) and greater sensitivity to inhibition by 2,4-diaminopyrimidines (section 9.4). This differential sensitivity forms the basis of the selective action of this group of drugs against *Plasmodium* and *Eimeria*. Why the drugs are not toxic to intact trypanosomes is not known. The origins of the active one-carbon transfer group have, in con-

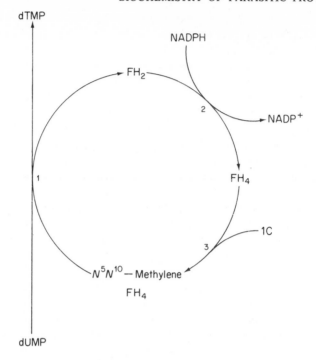

Figure 6.9 Thymidylate synthase cycle. Enzymes: 1, thymidylate synthase; 2, dihydrofolate reductase; 3, methyltransferase. FH_2, dihydrofolate; FH_4, tetrahydrofolate; 1C, one-carbon unit; other abbreviations as in figure 6.7.

trast, not been well studied, although in *Plasmodium* it appears to arise from the conversion of serine to glycine by the enzyme serine hydroxymethyltransferase. This enzyme has been isolated recently from *P. lophurae* and shown to have a molecular weight of 68 000 daltons, much lower than that of the other enzymes in the thymidylate synthase cycle.

The origin of the cofactor tetrahydrofolate has been investigated, especially in *Plasmodium*. This genus appears to synthesise tetrahydrofolate *de novo* from guanosine triphosphate (GTP) in a manner similar to that in bacteria (see figure 9.12). In contrast, mammalian cells salvage it from folic acid, a dietary vitamin, using the enzyme folate reductase. The different origins of tetrahydrofolate in *Plasmodium* and mammals is thought to be the basis of the selective action of sulphonamide drugs in malaria (section 9.4). The sensitivity of *Eimeria* and *Toxoplasma* to sulphonamides is taken to imply that pathways exist in these two genera similar to that found in *Plasmodium*. The situation in *Trypanosoma* is unclear, and there is no information about tetrahydrofolate synthesis in *Leishmania*, *Trichomonas*, *Entamoeba*, *Babesia* and *Theileria*.

Besides synthesis *de novo* from dUMP, dTMP can be salvaged also from thymine and thymidine (figure 6.7). Thymidine salvage is common both in parasitic protozoa (including *Trypanosoma*—except *T. vivax*—but not

Eimeria and *Plasmodium*) and in mammalian cells, but there are no well-documented cases of thymine salvage in parasitic protozoa. The mechanisms of regulation of the production of dTMP *de novo* and by salvage and particularly the control of the balance between these two pathways is not understood at present in any genus of parasitic protozoa.

6.5 Nucleic Acid Synthesis

DNA replication in mammalian cells is mediated by DNA polymerases which are probably membrane-bound. All four deoxyribonucleoside triphosphates (that is, dATP, dGTP, dTTP and dCTP) and a DNA template are required for their activity. The reaction is thought to be effectively irreversible because of the presence in cells of pyrophosphate phosphatase activity which rapidly breaks down one of the products, pyrophosphate, to inorganic orthophosphate. Synthesis is not continuous throughout the cell cycle, but is confined to a definite phase (S-phase) which is separated from nuclear and cell division by G_1 and G_2 phases. Other DNA polymerases are thought to be involved in various DNA repair mechanisms. RNA synthesis (that is, DNA transcription) in mammalian cells is mediated by RNA polymerases which require all four ribonucleoside triphosphates (ATP, GTP, UTP and CTP) and a DNA template for their activity. This reaction too is effectively irreversible.

There is no detailed information on either DNA or RNA polymerases in any genus of parasitic protozoan, although a preliminary account of two RNA polymerases from *Crithidia* has appeared. The kinetoplast DNA (section 6.7) of some species of trypanosome has been transcribed *in vitro* with bacterial RNA polymerases. The periodicity of DNA synthesis has been studied using [^{3}H]-thymidine and microautoradiography in *Crithidia* and some species of *Trypanosoma*. Both nuclear and kinetoplast DNA synthesis show a typical eukaryotic pattern of periodicity, with S-periods in most cases being almost or actually coincident. *T. equiperdum* is a possible exception; kinetoplast DNA synthesis appears to occur much earlier than nuclear DNA synthesis. DNA repair has not been studied.

6.6 DNA

The DNA normally present in mammalian cells is double stranded with the two strands running in opposite directions in the form of a double helix and with the nitrogen bases of the two strands paired such that adenine is associated with thymine and guanine with cytosine. The order of nitrogen bases along the helix is the key to the genetic information contained within that DNA. The base composition of DNA is expressed as the G + C content (that is, the percentage of guanine and cytosine residues in the DNA relative to the total number of residues). This can be determined experimentally, not only by direct chemical analysis on purified DNA but also from the buoyant density of that DNA in an isopycnic caesium chloride gradient or from the melting temperature (T_m) of the DNA in standard conditions (see Adams *et al.*, 1976). DNAs containing similar genetic information have similar base compositions; those containing different genetic information tend to have different base compositions, though not necessarily so, since base compositions reflect only on the total number of guanine and cytosine residues, not their order along the double helix.

The bulk of the DNA in mammalian cells is in the nucleus where it is associated with basic proteins called histones. There are about 5 pg of DNA in each cell and it is made up of long linear double-stranded molecules with very few unusual bases and with a base composition of about 40 per cent G + C (table 6.1). In addition to this nuclear DNA, there is usually a small quantity of DNA in the mitochondria of mammalian cells. This is composed of small circular double-stranded molecules with a contour length of about 5 μm (molecular weight about 10×10^6 daltons). It is believed that this DNA contains the genetic information for the synthesis of ribosomal (rRNA) and some transfer (tRNA) RNAs (section 6.8) and also a small number of proteins, some of which subsequently become associated with the membranes of the mitochondria.

All parasitic protozoa examined so far contain DNA, though the amount present, with the exception of *Eimeria* oocysts, which are possibly a special case since they already contain all the DNA for the eight sporozoites, is an order of magnitude less than that in mammalian cells (table 6.1). There is no information so far for *Theileria*. The base compositions of the major components, as determined from buoyant densities in CsCl and T_m, vary widely, both between and within genera. Thus the lowest so far recorded is 18 per cent G + C for *Plasmodium gallinaceum* and the highest 61 per cent for *Leishmania tropica*. Within the genus *Plasmodium*, there is a spread from 18 to 37 per cent G + C. It has been suggested that such differences within genera, which have also been noted, though to a lesser extent, with *Trypanosoma* and *Leishmania*, might be useful taxonomic parameters and they have indeed been used in this way in *Leishmania*. The base compositions obtained from buoyant density and T_m measurements do not agree in *Entamoeba*. The former suggest a base composition of 32 per cent G + C; the latter 22 per cent G + C. Thus it is probable that the DNA contains some unusual bases. In most genera, there is evidence from Feulgen staining and microautoradiography (though not from analysis of pure preparations of nuclei) for the main DNA components being present in the nuclei. It is also likely that the large satellite component (43 per cent G + C) of the *T. brucei* group trypanosomes, which comprises 38 per cent of the DNA, is of nuclear origin. A nuclear satellite DNA present at such high proportions is unique to this group of organisms. In *Trypanosoma*, there is evidence for the association of nuclear DNA with histones.

In *Trypanosoma*, *Crithidia*, *Leishmania*, *Trichomonas* and *Plasmodium*, there are small DNA satellites of different base composition to those of the main nuclear components. In *Plasmodium*, it has been shown recently in avian species that this satellite is associated with the mitochondria of the parasite and

Table 6.1 Properties of DNA from parasitic protozoa

Species	Amount per cell (pg)	Base composition (% G + C)		% Extranuclear
		Nuclear	Extranuclear	
Trypanosoma brucei	0.10	48 + 43	31	7*
Trypanosoma rhodesiense	0.10	48 + 43	30	7
Trypanosoma gambiense	0.10	48 + 43	30	7
Trypanosoma evansi	0.10	48 + 43	29	7
Trypanosoma equiperdum	0.10	48 + 43	33	7
Trypanosoma vivax	—	54	36	8
Trypanosoma congolense	—	49	37	8
Trypanosoma cruzi	0.17	50	39	25
Trypanosoma lewisi	—	47	39	—
Leishmania donovani	—	59	45	10
Leishmania tropica	—	61	47	10
Leishmania braziliensis	—	58	42	10
Trichomonas vaginalis	0.53	29	+	—
Trichomonas gallinae	0.40	34	+	—
Entamoeba histolytica	0.45	32–22†	—	—
Eimeria tenella	5.8	23–41‡	—	—
Toxoplasma gondii	0.10	52	—	—
Plasmodium falciparum	—	37	19	1
Plasmodium knowlesi	0.15	37	19	1
Plasmodium berghei	0.30	24	18	—
Plasmodium vinckei	—	24	—	—
Plasmodium gallinaceum	—	18	—	—
Plasmodium lophurae	0.13	20	19	—
Babesia rodhaini	—	36	—	—
Babesia microti	—	39	—	—
Mammalian cell	5.00	40	40	0.1

—, no information yet available; +, known to exist but base composition not yet determined; * blood stages—more in culture forms, see text; † calculations from T_m and density do not agree, see text; ‡ base composition wrongly stated in the literature.

that it is circular in form with a contour length of about 10 μm (molecular weight 21 × 10⁶ daltons). Thus it is very similar to, although somewhat larger than, mammalian mitochondrial DNA and probably serves a similar function. It is likely that similar components occur in the mitochondria of *Eimeria* and *Toxoplasma*, though as yet, there is no experimental evidence for this. In *Trichomonas*, there is some evidence that the hydrogenosomes (section 4.5) also contain a small amount of circular DNA of 3 μm contour length (molecular weight about 6 × 10⁶ daltons) although its base composition has yet to be estimated. There is no evidence about extranuclear DNA in *Entamoeba*, *Babesia* and *Theileria*.

6.7 Kinetoplast DNA

In *Trypanosoma*, *Crithidia* and *Leishmania*, the extranuclear DNA is associated with the kinetoplast. A proliferation of the membranes associated with this organelle into tubular-like structures makes up the mitochondrial-equivalent of these genera, so that it is generally accepted that kinetoplast DNA is the equivalent of mitochondrial DNA of other eukaryotic cells. The amounts of kinetoplast DNA in different species are given in table 6.1. In *T. brucei*, but not *T. cruzi*, there is an amplification (up to eight times) of the kinetoplast DNA in culture forms which may be connected with the metabolic changes described in section 5.2. A special feature of kinetoplast DNA is its ability to band rapidly in isopycnic CsCl gradients—it can be detected within one hour of the start of centrifugation and long before the gradient has come into equilibrium. Once it was realised that this feature was at least in part the result of the kinetoplast DNA within each kinetoplast being aggregated into a single network with a high sedimentation coefficient (S value) (molecular weight up to 6×10^{10} daltons), methods were evolved for its rapid purification from cell homogenates by a combination of differential and preparative CsCl or sucrose gradient centrifugation. Physico-chemical measurements on purified preparations of networks have established that each is equivalent to the DNA in one kinetoplast and that the networks are composed of DNA and very little else. In absolute amounts, there is about 0.01 pg of DNA in each. Constraints imposed on the DNA by its organisation into networks render it unmeltable by heating under standard conditions. After disruption by sonication, a conventional melting curve is obtained. Base compositions calculated from these curves agree with those calculated from data of buoyant density in CsCl gradients. This suggests that there are few unusual bases in kinetoplast DNA, a conclusion that has been confirmed recently by direct chemical analysis.

Visualisation in the electron microscope of these purified networks has been achieved by various modifications of the Kleinschmidt technique (figure 6.10). Early studies led to suggestions that the networks were composed of a mass of catenated (interlocked) double-stranded minicircles (so called because of their small size) and longer linear pieces. More recent observations have indicated that these linear pieces are in fact parts of larger maxicircles which are catenated into the minicircle network. Some of them appear to project from the network, thus forming the edge-loops seen in electron micrographs. The proportion of maxicircle DNA in the networks (in terms of mass) appears to range from about 3–5 per cent (*Crithidia*) to 20 per cent (*T. brucei*).

Minicircles vary in contour length from about 0.3 μm (molecular weight 0.5 $\times$ 10^6 daltons) in *Trypanosoma* and *Leishmania* to about 0.8 μm (1.6 $\times$ 10^6 daltons) in *Crithidia*. There are about 10 000 in each network. Kinetic complexity measurements made during thermal renaturation experiments and investigations with restriction endonucleases have indicated that within one network they are similar but not identical in size and sequence. This heterogeneity appears to be higher in *T. brucei* than it is in *T. cruzi* and

Crithidia. Maxicircles vary in contour length from about 6–7 μm (molecular weight 12–13 $\times$ 10^6 daltons) in *Trypanosoma* to about 12 μm (molecular weight 22 $\times$ 10^6 daltons) in *Crithidia.* They are not simply multimers of minicircles but, rather, they contain a non-repeated sequence of DNA. There are probably only 50–100 present in each network in *Crithidia.* All within a network appear identical in size and sequence, though there is some disagreement on this point.

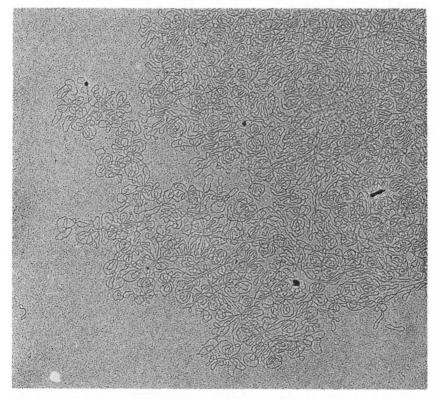

Figure 6.10 Electron micrograph of part of a kinetoplast DNA network from a blood trypomastigote form of *Trypanosoma cruzi* ($\times$ 30 000).

Studies of kinetoplast DNA replication in which organisms have been pulsed with [^{3}H]-thymidine and then chased with non-radioactive thymidine for different times before network isolation, have shown that label first appears in the outer regions of the networks but that subsequently it moves inwards and spreads throughout the entire structure. This suggests that the whole of the network is replicated during each cell cycle. Minicircles in replication have been described in *T. cruzi*; maxicircles in replication have yet to be seen.

The transcription of kinetoplast DNA in minicircles but not maxicircles has been carried out *in vitro* using bacterial RNA polymerase. It has been shown that the polymerase binds to AT-rich regions of the DNA and that there are four of these in each minicircle. RNA species with values of 9S and 12S (molecular weights of 0.2×10^6 and 0.46×10^6 daltons) have been isolated as products of this transcription. RNA species with the same S values have also been isolated from highly purified preparations of kinetoplasts. Their significance is not clear at present, although it has been suggested that they represent stable kinetoplast messenger RNAs (mRNAs).

The function of kinetoplast DNA has been the subject of much speculation and little fact. It is now generally agreed, however, that the maxicircles represent the true mitochondrial DNA and that, as in *Plasmodium* and mammalian cells, this contains genetic information for the synthesis of rRNA, some tRNAs and a few proteins. However, there is no experimental evidence as yet for this conclusion. It is also now generally agreed that the minicircles are not transcribed and translated. One possibility is that, instead, they have a structural role, ensuring an even split of maxicircle DNA during kinetoplast and mitochondrial division, a task of particular importance in organisms with single mitochondria.

Maxicircle kinetoplast DNA is probably only transcribed and translated in the insect vector forms of the *T. brucei* group trypanosomes. In these species, the TCA cycle enzymes and the cytochrome system are repressed in the mammalian stages and the mitochondria contain only very few cristae (section 4.3). Support for this idea came from observations that *T. evansi*, which appears to lack a maxicircle kinetoplast DNA component and dyskinetoplastic species such as *T. equinum*, in which the kinetoplast DNA is not in the form of a network, can survive in animals as long as they are transmitted mechanically, but are unable to infect tsetse flies, in the gut of which transformation to procyclic forms with developed mitochondria must occur (chapter 5).

6.8 RNA

In mammalian cells, three functional types of RNA exist. Messenger RNA occurs in the smallest quantity and is concerned with the transfer of genetic information from the nucleus to the polyribosomes of the cytoplasm where protein synthesis occurs. Transfer RNAs, of which at least one species exists for each amino acid, are concerned with the assembly of amino acids in the correct order during translation of mRNA on the ribosome. Ribosomal RNA occurs in the greatest quantity. At least four components exist—23S, 18S, 7S and 5S—the first three being derived from the same 45S precursor rRNA (table 6.2). They combine with certain proteins to make up the ribosomes on which protein synthesis occurs (chapter 7).

In parasitic protozoa, it is assumed that mRNA is formed by transcription of nuclear DNA, but there is no direct evidence for this and the half life of mRNA (which is relatively short in bacteria and long in mammalian cells) is not known. Only in *Trypanosoma* and *Plasmodium* is there any evidence of a tRNA system. Ribosomal RNA has, however, been isolated from *Trypanosoma*, *Crithidia*, *Entamoeba*, *Plasmodium* and *Toxoplasma* (table

6.2). Note the apparent absence of a 7S component (this is absent also from bacteria) and the S values of the other components compared to those of mammalian and bacterial rRNAs. Unusual bases (for example, O-2′-methylinosine) have been detected in *Crithidia* rRNA and there is some evidence for an uneven distribution of adenine and uracil resides in *Plasmodium* rRNA.

Table 6.2 Ribosomal RNA in parasitic protozoa

Genus	Ribosomal RNA components*			
	1	2	3	4
Bacteria	23 (1.1)	16 (0.6)	—	5
Trypanosoma cruzi	26 (1.45)	21 (0.92)	—	+
Crithidia	25 (1.30)	20 (0.83)	—	—
Entamoeba	28 (1.8)	18 (0.7)	—	—
Toxoplasma	24 (1.2)	19 (0.77)	—	4.5
Plasmodium	24 (1.2)	17 (0.65)	—	3.8
Mammal	28 (1.8)	18 (0.7)	7	5

—, not detected so far; +, present but S value not determined.
* Figures are the S values plus, in parentheses, the approximate molecular weights ($\times$ 10^6 daltons).

6.9 Nucleic Acid Catabolism

In the lysosomes of mammalian cells, there are nucleases such as deoxyribonuclease (DNAase) and ribonuclease (RNAase) which degrade DNA and RNA to nucleotides. Other enzymes, especially phosphatases and phosphorylases, further degrade the nucleotides to nucleosides and free bases (figures 6.3 and 6.7).

It is assumed that similar nucleases occur in parasitic protozoa especially since lysosomes have been detected in a number of species. However, none has been examined in any detail, except the DNAase of *Entamoeba*, though there is new preliminary evidence for high activity in *Trypanosoma cruzi*. There is evidence in *Crithidia* and the culture epimastigote forms of *Trypanosoma cruzi* that dTMP synthesis is not properly regulated and that the excess not required for DNA synthesis is degraded to thymine by the combined function of a thymidylate phosphatase and a thymidine phosphorylase activity (figure 6.7). Nucleoside hydrolases have also been described in some members of the *Trypanosoma*.

6.10 General Conclusions

It is clear from the previous sections that there is a considerable body of evidence available now about nucleic acid metabolism in parasitic protozoa, although there are still enormous gaps in our knowledge. Thus, while there have been many studies which have established that most parasitic protozoa can synthesise pyrimidines but not purines *de novo*, the polymerisation of these compounds into nucleic acids has hardly been investigated at all. One reason for a lack of information on nucleic acid synthesis is undoubtedly the problem of the availability of material.

In the areas where results have been obtained, a number of structures, pathways and enzymes have been uncovered which are actual or possible targets in selective drug therapy. These include dependence on preformed purines, dihydroorotate hydroxylase and dihydroorotate dehydrogenase, a tetrahydrofolate synthetic pathway, dihydrofolate reductase, thymidylate synthase, kinetoplast DNA and ribosomal RNA.

6.11 Further Reading

Adams, R. L. P., Burdon, R. H., Campbell, A. M. and Smell'e, R. M. S. (1976). *Davidson's The Biochemistry of Nucleic Acids*. Eighth edition, Chapman and Hall and Science Paperbacks, London

Borst, P. and Fairlamb, A. H. (1976). DNA of parasites with special reference to kinetoplast DNA. In *Biochemistry of Parasites and Host–Parasite Relationships* (ed. V. Van den Bossche), Elsevier-North Holland Biomedical Press, Amsterdam, pp. 169–191

Jaffe, J. J. and Gutteridge, W. E. (1974). Purine and pyrimidine metabolism in protozoa. *Actualités Protozoologiques*, **1**, 23–35

7 Protein Metabolism

7.1 Introduction

Proteins are fundamental to all aspects of cell metabolism and are found throughout every cell. They normally constitute more than 50 per cent of its dry weight, and are the most important agents for the expression of the genetic material of the cell.

Proteins differ greatly in structure and function. The largest class are the enzymes, of which nearly 2000 are known, which are highly specific catalysts of the cell. Other classes include the hormones, contractile, structural, storage and transport proteins. The common characteristic of cell proteins is that they consist of L-α-amino acids, linked in long chains by peptide bonds (figure 7.1). A polymer of amino acids is called a polypeptide, one or more chains of which make up a protein. The molecular weight of proteins varies from a few thousand to greater than 1 million daltons. However, all proteins are composed basically of the same 20 amino acids (table 7.1). The sequence of amino acids characteristic for each polypeptide chain is called the primary structure and ultimately determines the three-dimensional structure of the protein. This primary structure is itself determined by the sequence of bases in the DNA of the cell. Some proteins contain only amino acids and are called simple proteins, whereas others have a metal or organic prosthetic group and are known as conjugated proteins (for example, glycoproteins, lipoproteins). Some proteins are closely associated with the membraneous structures of the cell and are described as particulate, whereas others are soluble.

The proteins of mammals have been studied extensively. This is not the place to attempt to summarise our present understanding of protein structure and function; it is too large a field. In comparison, the study of the proteins of parasitic protozoa, their structure, synthesis, activity and function, is still very much in its infancy. This chapter will deal only with those aspects of proteins that have been investigated in parasitic protozoa. Enzymes are proteins, and many enzymes have been found in parasitic protozoa. However, in most cases these are discussed in other chapters, when the reaction which they catalyse is considered.

7.2 Sources of Amino Acids

All cells require amino acids essential for the synthesis of proteins. Mammals

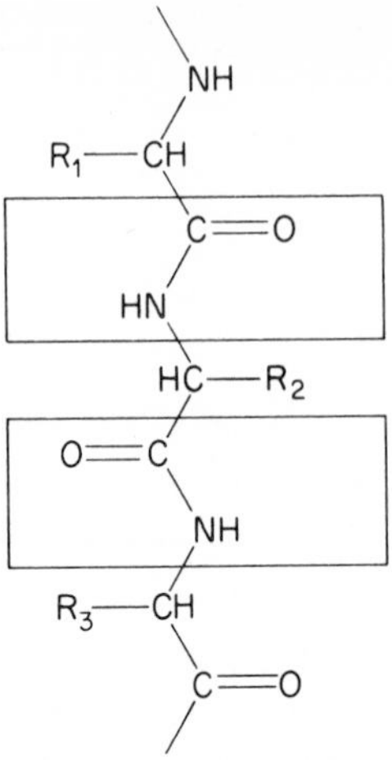

Figure 7.1 General structure of a protein. The peptide bonds are boxed. R_1, R_2, R_3 correspond to the R groups of different amino acids.

Table 7.1 The common amino acids, classified according to the polarity of their R group

Non-polar or hydrophobic	Neutral (uncharged) polar	Negatively charged	Positively charged
Alanine	Asparagine	Aspartate	Arginine*
Isoleucine*	Cysteine	Glutamate	Histidine*
Leucine*	Glutamine		Lysine*
Methionine*	Glycine		
Phenylalanine*	Serine		
Proline	Threonine*		
Tryptophan*	Tyrosine		
Valine*			

* Essential amino acid of man

are unable to synthesise, from any precursor, ten of the common amino acids (table 7.1). These must be obtained from their food, and are called the essential amino acids. Mammals synthesise the remainder, either *de novo* or from the essential amino acids. Ammonium ions are used in these syntheses; other inorganic nitrogen molecules are not incorporated.

Parasitic protozoa require the same 20 amino acids for their proteins. As in mammals these can be obtained from the environment or by synthesis. Our knowledge of the precise dietary amino acid requirements of parasitic protozoa is very poor, but in all cases it seems that some are required preformed. Table 7.2 lists those known to be utilised by different species, although the ability to use an amino acid does not indicate it is an essential requirement.

Table 7.2 Amino acids taken up by some parasitic protozoa

	Alanine	Arginine	Asparagine	Aspartic acid	Cysteine	Glutamic acid	Glutamine	Glycine	Histidine	Isoleucine	Leucine	Lysine	Methionine	Phenylalanine	Proline	Serine	Threonine	Tryptophan	Tyrosine	Valine	Method
Trypanosoma brucei C	−	−	−	−	−	+	+	−	−	−	+	−	−	−	+	+	+	−	−	−	D
Trypanosoma lewisi B		+	+	+		+															O
Trypanosoma cruzi C	−	+	+	+		+	+	−	+	+	+	+	+	+	+	+	+		+	+	D
Leishmania tarentolae C	+						+			+	+		+	+	+	+	+	+	+	+	E
Crithidia fasciculata C	+								+	+	+	+	+	+				+	+	+	E
Trichomonas foetus C	+							+	+	+	+	+	+	+	+	+				+	E
Plasmodium knowlesi B			+						+	+	+		+								L
Plasmodium lophurae B	+	+	+		+	+	+	+	+	+	+					+	+	+		+	L
Plasmodium relictum S		+	+			+	+	+	+				+			+	+				D

B, blood stages; C, culture stages; S, oocysts; E, essential in culture; O, stimulates oxygen consumption; L, labelled compound incorporated; D, chemical determination; +, utilised; −, not utilised

Some details of the mechanisms by which parasitic protozoa satisfy their amino acid requirement, and the source of these amino acids, is known.

7.2.1 Proteins

All parasitic protozoa possess the ability to hydrolyse proteins within the cell, for this is an essential requirement for the control, by inactivation of enzymes, of growth and development. This autolysis yields amino acids which are available for protein synthesis. Some parasitic protozoa also use exogenous protein as a source of amino acids. In most cases these proteins are ingested by the protozoan, either through a cytostome, or by phagocytosis or pinocytosis, into food vacuoles. The protein then is hydrolysed in the vacuole by the lysosomal enzymes (for example, cathepsins and other peptidases) which presumably are released into the vacuole when lysosomes fuse with it. Some parasitic protozoa, however, apparently excrete proteases which hydrolyse host proteins outside of the parasite. The amino acids produced can then be absorbed. Careful investigations are required to determine the localisation of any enzyme, and in most cases these have not been carried out on the proteases of parasitic protozoa. However, proteases have been identified and in some cases characterised.

The protease activity of the erythrocytic stages of *Plasmodium* species have been shown to differ from the host cell enzymes in the optimum pH for activity and the sensitivity to inhibitors. The main enzyme is an acid protease similar to mammalian cathepsin D. Malaria parasites obtain amino acids by digestion of the host cell haemoglobin. It has been suggested that protease inhibitors may be of value in inhibiting the growth of the malaria parasite by selectively acting on a process that is essential to it but not, at least over a short period, to the cells of the host. There is a suggestion that different species of *Plasmodium* possess proteases specific to the haemoglobin of their host, which possibly could account, in part, for the very great host specificity of these parasites. Several proteases with different specificities have been identified in *Entamoeba histolytica*. These include enzymes similar to mammalian trypsin and pepsin as well as a gelatinase, a casease and a glutaminase. The localisations of the enzymes within the cell have not been established fully. A lysosome-like system exists, but must differ from the equivalent system in mammalian cells as *Entamoeba* species apparently possess neither Golgi apparatus nor a functional endoplasmic reticulum. There is evidence for the presence of specialised surface-active lysosomes; membrane-bound acid hydrolases are present. These may be important in tissue invasion by *Entamoeba*. Several proteases have been reported to be present in blood-stream trypanosomes, including cathepsin, carboxypeptidase, aminopeptidase and dipeptidase. There is evidence for the existence of lysosomes, but also many hydrolases are apparently associated with the flagellar pocket and may be concerned with extracellular digestion prior to pinocytosis. This excretion of hydrolytic enzymes may be responsible in part for the pathogenicity of the disease. Protease activity in *Trichomonas foetus* is associated with lysosome-like organelles.

7.2.2 Exogenous amino acids

Amino acids enter mammalian cells variously by active transport (at least five specific transport mechanisms have been found), passive mediated transport, simple diffusion and a novel cyclic mechanism involving group translocation.

Parasitic protozoa will utilise free amino acids from their environment (table 7.2), although the mechanisms of entry of these amino acids into the cells have been investigated very little. Amino acids enter trypanosomes both by simple diffusion and by mediated processes involving several sites, the specificites of which are unusual and different from those of mammalian cells. For example, in *Trypanosoma brucei* four mediated transport sites have been characterised. The N_1 site is specific for serine, threonine and alanine, is quite distinct from the other sites and can be selectively inhibited by a glycine analogue. The N_2 site is specific for neutral amino acids of aromatic character or with hydrocarbon side chains. There are also separate sites transporting acidic amino acids and basic amino acids. The entry of amino acids into the intraerythrocytic malaria parasite is made easier by the increase in permeability of the red cell membrane induced by the parasite (section 4.8). This change facilitates entry of some amino acids, which are transported into uninfected erythrocytes by energy-linked mediated processes, by simple diffusion while others, although still requiring mediation, enter more easily than into normal red blood cells.

7.2.3 Synthesis *de novo* and interconversion of amino acids

Mammals synthesise the ten non-essential amino acids *de novo* from intermediates of carbohydrate metabolism, or from one or more of the essential amino acids. Although these pathways exist, it is clear that in many instances the requirement for an amino acid will be satisfied by the content of the food.

Few thorough studies of amino acid synthesis in parasitic protozoa have been carried out. However, the non-utilisation of some exogenously supplied amino acids indicates that some pathways exist, a few of which have been reported. Blood-stream stages of *Trypanosoma rhodesiense* can synthesise alanine, glycine, serine, aspartic acid and glutamic acid from carbohydrate, whereas *T. cruzi* culture forms can produce alanine, aspartate, glutamate,

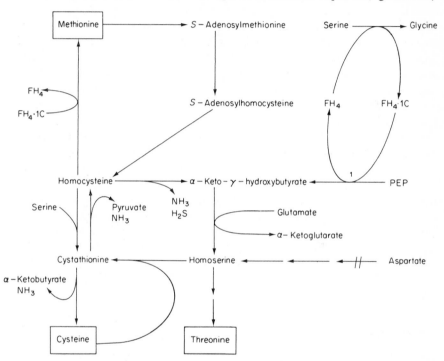

Figure 7.2 The pathways involved in the synthesis of threonine and methionine from PEP, and the conversion of methionine to threonine and cysteine in *Crithidia fasciculata*. Enzyme: PEP tetrahydrofolate hydroxymethyltransferase. FH$_4$, tetrahydrofolate; ⫫, enzyme absent.

glycine, cysteine and threonine from serine. Culture forms of *T. brucei* can convert proline to alanine and aspartate, although this is considered to be mainly a catabolic pathway (section 5.2). The synthesis of threonine and methionine in *Crithidia fasciculata* has been studied in some detail. Mammals normally derive threonine from aspartate, but the first enzyme of this pathway, aspartate kinase, is absent from *C. fasciculata*. Instead this protozoan can produce threonine from methionine or, in its absence, from cysteine if high levels of folate are present. The pathways involved in the synthesis of both threonine and methionine in these conditions are shown in figure 7.2. The key compound is

α-keto-γ-hydroxylbutyrate. It is formed from phosphoenolpyruvate and the β-carbon of serine in a reaction requiring high levels of the one-carbon transfer compound tetrahydrofolate. The enzyme catalysing this novel reaction, which at present is unique to *C. fasciculata*, has been named phosphoenolpyruvate-tetrahydrofolate-hydroxymethyltransferase. If a similar pathway was found to exist in pathogenic trypanosomes, it clearly would be a possible target for chemotherapy. *Crithidia oncopelti* will grow with methionine as the only dietary amino acid. It can synthesise the other amino acids required. It is probable, however, that much of the synthetic ability is due to the 'bipolar body' (see appendix C).

Less is known about other parasitic protozoa. Proline plays an important role in the metabolism of *Leishmania* (section 4.4) and can serve as the major energy source. However, in the absence of this amino acid *L. tarentolae* will continue to grow, synthesising proline, required for protein synthesis, from arginine. The synthetic pathway involves ornithine, is not regulated by proline, but is yet to be elucidated fully. Surprisingly, glutamate, the catabolic breakdown product of proline, cannot act as a precursor of the imino acid. Nor can other amino acids. Species variation among the genus *Leishmania* is demonstrated by the finding that arginine is an essential amino acid for *L. tarentolae* and cannot be synthesised from proline, whereas *L. donovani* can carry out this conversion. The malaria parasite *Plasmodium knowlesi* can synthesise alanine, aspartic acid and glutamic acid (but seemingly no other amino acids) from carbohydrate, cysteine from methionine, and methionine from serine and homocysteine. *Entamoeba histolytica* can similarly convert carbohydrate to alanine, aspartic acid and glutamic acid only. Many aminotransferases possibly involved in the inter-conversion of amino acids have been reported. However, it is probable that these are more important in the catabolism of amino acids (section 7.4).

7.3 Protein Synthesis

The mechanisms by which proteins are synthesised in mammalian cells and bacteria have been studied extensively. The mechanisms are similar in the two groups but differ in detail. These differences constitute the basis of activity of many antibacterial drugs. In contrast, protein synthesis in parasitic protozoa has been investigated very little, despite its potential importance. Our present knowledge indicates that the mechanisms in parasitic protozoa are more similar to those operative in mammalian cells than those of bacteria.

Protein synthesis in mammalian cells is complex and will not be described in detail here. The sequence of events is as follows. Nuclear DNA is transcribed to produce messenger RNA (mRNA), a process that occurs in the nucleus and is catalysed by the enzyme DNA-primed RNA polymerase. The mRNA migrates to the cytoplasm, where several ribosomes attach to it, the nucleic acid passing between the two subunits of the ribosome. Ribosomes are the particles where the genetic code of the mRNA is read and translated into a polypeptide sequence. Therefore they are the actual sites of protein synthesis, and will be described in more detail later. The ribosomes move along the mRNA and amino acids, for which each successive triplet of bases of the mRNA codes, are lined up and covalently joined in peptide bonds. The amino acids are transported to ribosomes by a type of RNA—transfer RNA (tRNA)—of which there are many kinds, each specific for one amino acid. Each possesses a base triplet complementary to the mRNA triplet that codes for the particular amino acid. The amino acid is joined to the tRNA in a reaction catalysed by the enzyme amino acid tRNA synthetase. When a polypeptide chain is completed it is released by the ribosome.

Mammalian cells also synthesise proteins in their mitochondria with enzymes specific to the mitochondrial process and distinct from those of the cytoplasm. This process and its sensitivity to drugs is more similar to that occurring in prokaryotes than that in the cytoplasm of mammalian cells. Protein synthesis in bacteria and mitochondria are typically inhibited by choloramphenicol, whereas mammalian cytoplasmic protein synthesis is inhibited by cycloheximide.

Prokaryotic and mammalian cytoplasmic ribosomes, although basically similar, differ characteristically. These differences are intimately related to the differing drug sensitivities. All ribosomes contain two subunits, of differing size, and consist of protein and ribosomal RNA (rRNA) in the approximate ratio 1 : 2. Mammalian cytoplasmic ribosomes are bigger, however, the whole particle being approximately 20 nm in diameter, with a molecular weight of 4×10^6 daltons and an S value of 80. The two subunits typically are 60S and 40S. This compares with bacterial ribosomes of 70S, with subunits of 50S and 30S. Lower eukaryotes possess cytoplasmic ribosomes in the size range of 73–80S. All mitochondrial ribosomes, irrespective of source, are sensitive to most of the inhibitors of bacterial protein synthesis but insensitive to cycloheximide. However, they differ in size; there are apparently two size ranges. Primitive eukaryotes such as fungi and protozoa (most data are from free-living protozoa) possess mitochondrial ribosomes of size 70–80S, whereas in higher eukaryotes (from sharks to man) they are between 50S and 60S, typically with subunits of approximately 45S and 35S.

Our knowledge of protein synthesis in parasitic protozoa is slight even though the trypanocide, antrycide and the antimalaria drugs, minocycline and clindamycin appear to block this area of metabolism (section 9.6). That it occurs is obvious, and it has been demonstrated many times. However, few details of the systems involved have been elucidated. Ribosomal structure has been studied to an extent.

Most work has been done with the trypanosomatids. Both cytoplasmic and mitochondrial protein synthesis occur, possibly even in blood-stream trypanosomes which possess acristate mitochondria. Mitochondrial protein synthesis probably, however, accounts for only about 3 per cent of the total synthesis. The initiation of polypeptide synthesis in the cytoplasm of *Crithidia* has been studied to an extent, the present evidence suggests that it is more similar to that occurring in the cytoplasm of other eukaryotic cells (involving methionyl-tRNA) than that in mitochondria and prokaryotes (involving N-formylmethionyl-tRNA). Cytoplasmic protein synthesis of trypanosomes is inhibited by cycloheximide, but not chloramphenicol. Mitochondrial synthesis is sensitive to chloramphenicol in most cases, although *Crithidia fasciculata* may be an exception. The cytoplasmic ribosomes are in the size range 83–88S, and, similar to mammalian ribosomes, require Mg^{2+} for activity. However, like prokaryotic ribosomes, they are more easily dissociated by lowering the Mg^{2+} concentration than mammalian ribosomes. Mitochondrial ribosomes of 72S have been reported.

Protein synthesis by cell-free extracts of intraerythrocytio stages of malaria parasites is inhibited by cycloheximide but not chloramphenicol and streptomycin and so protein synthesis in *Plasmodium* is probably similar to that in the cytoplasm of mammalian cells. *Plasmodium knowlesi* possess 80S cytoplasmic ribosomes, easily dissociated into 60Sand 40S subunits, which are superficially similar to those of mammalian cells. However, the rRNA has a G + C content of 37 per cent, which is similar to that of some free-living protozoa, but quite distinct from that reported for the rRNA of mammalian cells such as reticulocytes. It has been reported that amino acid tRNA synthetases of *Plasmodium* can use host cell tRNA, whereas the reverse is not the case. However, this awaits confirmation. Nothing is know of mitochondrial protein synthesis in malaria parasites.

Entamoeba species possess typical 80S eukaryote-type ribosomes that are synthesised during the trophozoite stage. Most of the ribosomes are free and not membrane bound; there is little if any endoplasmic reticulum present. In the late trophozoite the ribosomes form short helices, 40 nm in diameter, which aggregate in arrays. These aggregates form the chromatoid bodies of the cysts and contain 90 per cent of the ribosomes present in the cyst. They are considered to be ribosome stores for use in protein synthesis after excystation.

Clearly a great deal remains to be discovered about protein synthesis in parasitic protozoa. No details of the enzymes involved are yet available. As expected the systems are superficially similar to those present in other eukaryotes. Our present knowledge indicates that the main difference from mammals is that mitochondrial ribosomes of parasitic protozoa are rather larger. Whether there is an equivalent mechanistic difference which can be exploited and even whether inhibition of mitochondrial protein synthesis would be seriously harmful to the parasite are at present open questions.

7.4 Catabolism and other Metabolism of Amino Acids

The catabolism of amino acids can lead to the production of energy, and in some circumstances amino acids are an important energy source. However, the nitrogenous waste products that arise from the amine groups, in the catabolism of amino acids, must be excreted and in mammals this requires the utilisation of energy.

In mammals, amino acids serve as a major energy source only when present in quantities excess to the biosynthetic needs of the body, or when starvation is imminent, although the catabolism of endogenous amino acids, released when cellular proteins are hydrolysed, occurs continually with a concomitant production of energy. When used as energy sources, amino acids are deaminated and the remaining carbon skeleton is either converted to glucose or oxidised to CO_2 through the TCA cycle. In either case the entry of the carbon skeletons into the intermediary metabolic pathways is the same (figure 7.3) and

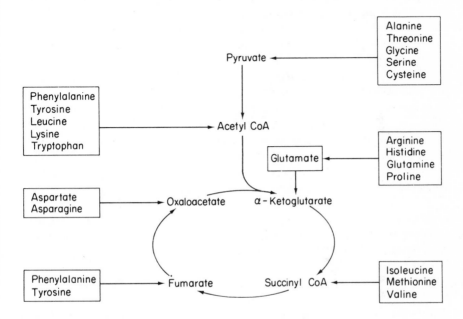

Figure 7.3 The points of entry into the TCA cycle of the carbon skeletons of amino acids. Amino acids are in boxes.

ultimately they are oxidised, releasing energy. When amino acids are oxidised in this way, the amine groups normally are removed by transamination to α-ketoglutarate, yielding glutamate. This undergoes deamination in a reaction catalysed by glutamate dehydrogenase

$$\text{glutamic acid} + NAD(P)^+ + H_2O \longrightarrow \alpha\text{-ketoglutarate} + NH_3 + NAD(P)H + H^+$$

Some L-amino acids and all D-amino acids are catabolised by oxidases present in the peroxisomes of liver and kidney cells. The net reaction is

$$\text{D-amino acid} + H_2O + O_2 \longrightarrow \alpha\text{-keto acid} + NH_3 + H_2O_2.$$

The ammonia produced is converted in the liver to urea, the final nitrogen excretary product of mammals, by a cycle of reactions known as the urea cycle

$$2NH_3 + CO_2 + 3ATP + 3H_2O \longrightarrow H_2N\underset{\text{(urea)}}{\overset{\overset{\displaystyle O}{\|}}{-C-}}NH_2 + 2ADP + AMP + 4P_i$$

Other vertebrates produce other nitrogenous excretions. The major form in which the nitrogen of amino acids is excreted in birds is uric acid, whereas freshwater fish excrete ammonia. Uric acid is also the end product of purine metabolism in primates and birds, whereas other mammals excrete allantoin. Pyrimidines are normally degraded to urea and ammonia.

Amino acid catabolism differs in degree and importance in different parasitic protozoa. It is possible that intraerythrocytic malaria parasites excrete amino acids; certainly there is a net production of amino acids by red blood cells infected with *Plasmodium berghei*. However, it is probable that this is the result of the release, but non-utilisation, of amino acids from the host cell haemoglobin. Most parasitic protozoa studied do catabolise amino acids. The culture forms of *Trypanosoma brucei* and several *Leishmania* species will utilise proline as a major energy source as described earlier (section 5.2). Other species, including *T. cruzi*, *T. lewisi*, *Trichomonas vaginalis* and avian *Plasmodium* species will use proteins or amino acids as energy sources if available in excess or if carbohydrate is lacking. Exogenous threonine is the major source of acetate units for lipid synthesis in *Trypanosoma brucei* and *T. cruzi* culture forms and in *Crithidia fasciculata* but not apparently in culture promastigote forms of *Leishmania donovani* and *L. braziliensis* (section 8.4). Many parasitic protozoa contain various aminotransferases, which are probably involved in amino acid catabolism.

The catabolism of the aromatic amino acids tyrosine, phenylalanine and tryptophan in trypanosomatids has been studied in detail. All are first deaminated by specific aminotransferases, but the α-keto acids produced, instead of undergoing further catabolism, are excreted as such or as closely related compounds. This is summarised in figure 7.4. Indole pyruvate has not

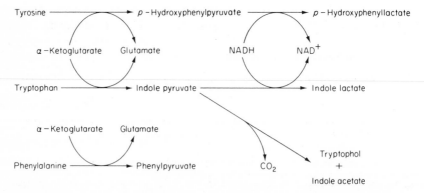

Figure 7.4 Catabolism of aromatic amino acids in blood-stream *Trypanosoma gambiense*.

been isolated, but is the presumed intermediate in the production of indole lactate (in the presence of NADH), tryptophol (indole ethanol) and indole acetate (in the absence of NADH). The enzyme systems involved in the production of tryptophol and indole acetate in blood-stream forms of *T. gambiense* possibly are particulate and differ from the analogous mammalian systems both in coenzyme requirements and sensitivity to inhibitors. However, they are yet to be fully characterised. Blood-stream forms of *T. gambiense* lack a lactate dehydrogenase, indicating that the enzyme responsible for the production of indole lactate and *p*-hydroxyphenyllactate may be specific for these reactions. Other Kinetoplastida also carry out some of these conversions. *Leishmania donovani* promastigotes apparently possess all three aminotransferases, and *C. fasciculata* contains tyrosine aminotransferase at levels twenty times greater than those found in rat liver. The role of these pathways in *T. gambiense* is unknown. Indole acetate is a plant growth hormone; a comparable function in trypanosomes is unlikely. A possible role in providing amino groups for keto acids and so synthesising amino acids needed for protein synthesis (for example, alanine, glutamate and aspartate) has been suggested. Certainly pyruvate, the chief end product of metabolism in blood-stream trypanosomes, can act as an amino acceptor in the aminotransferase reactions, as it can also in *L. donovani*. The production of the aromatic lactates may assist in regenerating NAD. These metabolic capabilities of the trypanosomes may contribute to the pathology of trypanosomiasis. For instance, utilisation of host tyrosine, phenylalanine and tryptophan may lead to a depletion of these amino acids in the host with a resultant lack of such important compounds as catecholamines and serotonin. Apparently, rats infected with *T. gambiense* convert tryptophan mainly to tryptophol and indole acetic acid, whereas mammals normally metabolise this amino acid by the serotonin pathway or the kynurenine pathway to nicotinamide. Such changes could result in behavioural depression and alterations to sleep and activity patterns, body temperature, blood pressure and glycogen and lipid metabolism. The metabolites of the pathways also may have a direct effect on the host; phenylpyruvate is known to inhibit adrenaline synthesis for instance.

Amino acid catabolism in other parasitic protozoa has been studied little. One important enzyme, glutamate dehydrogenase, has been shown to be present in several species. The enzymes fo the intraerythrocytic stages of *Plasmodium berghei* and *P. lophurae* and the epimastigote form of *T. cruzi* require NADP as coenzyme, and this reaction possibly is a major source of NADPH to the parasites. In the species which possess an active TCA cycle, the α-ketoglutarate produced in the reaction is also a source of energy.

The major nitrogenous excretory product of parasitic protozoa is ammonia. This is produced both by the glutamate dehydrogenase and D-amino acid oxidase reactions. The latter enzymes have been identified in intraerythrocytic stages of *P. berghei* and are probably quite widespread, if not particularly active. Other nitrogenous excretion products may occur. Several amino acids are excreted by intraerythrocytic malaria parasites, and alanine and glycine are excreted by *T. cruzi*, *T. brucei* and *L. donovani* growing in culture (chapter 5). Trypanosomes growing in culture also excrete thymine, due to an over-production of pyrimidines (section 6.9). The evidence suggests that no parasitic protozoan possesses an active urea cycle.

Amino acids are precursors of other important molecules as well as proteins. This aspect of metabolism has been ignored in the parasitic protozoa, except for some Kinetoplastida. *Crithidia fasciculata* can use tyrosine and dihydrophenylalanine to synthesize the hormone nerepinephrine although phenylalanine cannot be used in the absence of the enzyme phenylalanine-4-hydroxylase. The same protozoan also contains serotonin (5-hydroxytryptamine), a neurohormone and vasoconstrictor in vertebrates, which presumably it synthesises from tryptophan. In contrast *T. gambiense* cannot synthesise serotonin. The contribution of these two compounds to the cellular metabolism of *C. fasciculata* is unknown.

7.5 Occurrence and Function of some Proteins

Proteins are ubiquitous in occurrence and function. They are essential in all aspects of cellular metabolism, especially as enzymes. Enzyme proteins will be discussed in other sections of this book in most cases. Some proteins do not fit readily into any of the other sections, however, and these, together with a few interesting applications of proteins, will be considered in this section.

7.5.1 Membrane proteins

The protein contents of the membranes of parasitic protozoa have been little studied. The surface membrane of *Entamoeba invadens* contains nine polypeptides, whereas the cytoplasmic membranes contain sixteen polypeptides, which vary in molecular weight between 70 000 and less than 10 000 daltons. Mammalian membranes, in comparison, contain many more and larger proteins; however, the total amino acid content is similar. The surface membranes of *Entamoeba* species possess an unusual lipid content (chapter 8) and enzyme organisation (section 7.2) and are considered to be functionally quite distinct from those of both multicellular eukaryotes and other protozoa. A glycoprotein, molecular weight 30 000 daltons, consisting of 40 per cent carbohydrate and present in abundance in *Eimeria* oocyst cytoplasm, apparently is an important component of sporozoite membranes.

7.5.2 Proteins and antigenic variation

Antigenic variation was first studied in detail in African trypanosomiasis. During a chronic infection of *Trypanosoma brucei* in a mammalian host, the parasitaemia is not constant but shows a series of peaks and troughs. Immunological analyses showed that antibody was produced to the surface antigens of organisms in each population and this caused its decline but that each succeeding population had different surface antigens to the previous ones. Thus survival of the parasite during a chronic infection was a consequence of the ability of the organism to change its surface antigenicity. This phenomenon is referred to as antigenic variation. Originally, there was much discussion as to whether it was brought about by genotypic (that is mutation) or phenotypic variation in the population. Subsequent demonstration that it occurs in cloned strains and that the same antigenic variants often occur early in infection after vector passage led to its acceptance as a phenotypic phenomenon.

The biochemical basis of antigenic variation is now beginning to be un-

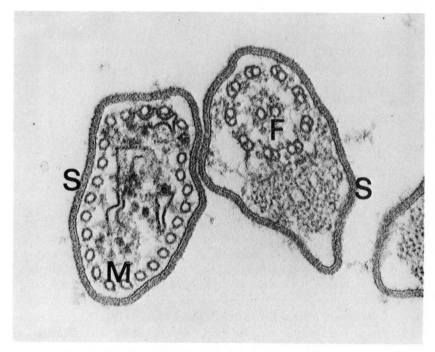

Figure 7.5 Electron micrograph of a section through the anterior end of a blood trypomastigote form of *Trypanosoma brucei* showing the surface coat around both the body and the flagellum (× 125 000). F, flagellum; M, subpellicular microtubules; S, surface coat.

derstood. Electron microscopic investigation of the surface structures of African trypanosomes have shown that the blood stages have a 15 nm-thick surface coat covering the entire body and flagellum (figure 7.5). This can be removed by washing or incubation with pronase or trypsin (both proteolytic enzymes) and disappears when organisms are cultured. In these circumstances, the surface charge is altered and the variant-specific antigenicity is lost. These observations suggested that the variant antigens are proteins of the surface coat and this was later confirmed in studies with ferritin-coated antibodies.

Formylmethioninesulphone-methyl-phosphate (FMSMP, figure 7.6) attaches specifically to proteins on the surface of cells. Its negative charge prevents it passing through biological membranes but it is a powerful N-acylating agent and thus readily adds covalently a formylmethionyl sulphone residue to free amino groups which abound on the surface of proteins (especially the ε-NH_2 group of lysine). Incubation of the blood stages of a cloned strain of *T. brucei* with [^{32}S]-FMSMP leads to the labelling of a single soluble protein which can easily be purified since it is excluded on columns of Sephadex G-25 and DEAE cellulose. It

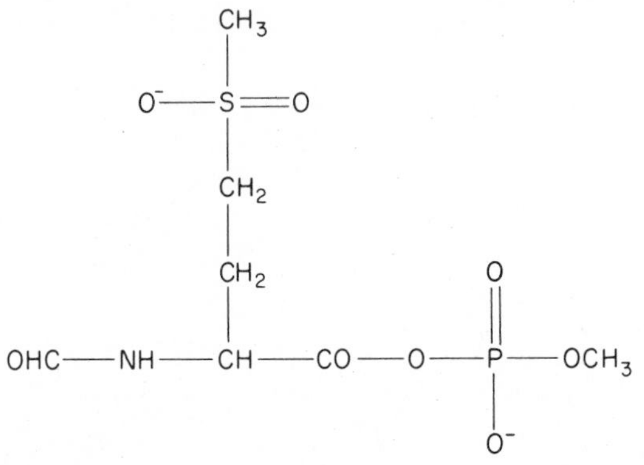

Figure 7.6 Chemical structure of FMSMP (formylmethioninesulphone-methyl-phosphate).

comprises about 10 per cent of the total proteins of the organism, which corresponds well with that calculated for the surface coat assuming it is homogeneous. It can be concluded therefore that the surface coat probably consists of a single protein. Characterisation of this protein showed that it was a glycoprotein with a molecular weight of 65 000 daltons (about 600 amino acid residues), although at present this is a controversial point, and a carbohydrate content of 3–6 per cent (mannose, galactose, glucosamine only). The same protein isolated from clones of other antigenic variants of the same strain of *T. brucei* had similar molecular weights and carbohydrate contents. Their isoelectric points (pI) and amino acid compositions, however, were quite distinct (table 7.3) and there is biochemical evidence that the differences are dispersed over more than half of the polypeptide chain. This is in agreement with immunological tests which have indicated no interaction between heterologous combinations of antiserum and antigens, suggesting that there are few areas of common structure in the variant glycoproteins studied so far. Thus, it is now thought that the surface coat of *T. brucei* is composed of a layer of homogeneous glycoprotein. Its structure varies markedly from clone to clone, conferring a unique immunological identity to each clone. It is not surprising therefore that, all too often, the host's immune system is unable to eradicate such an infection.

Preliminary observations on other species of the *T. brucei*-group of trypanosomes and on *T. congolense* and *T. vivax* suggest the presence of a surface coat similar ultrastructurally to that found in *T. brucei*. The blood stages

Table 7.3 Isoelectric point (pI) and amino acid composition* of principal surface glycoprotein from three clones of *T. brucei* taken from a chronically infected rabbit

pI or Amino Acid	Days in rabbit before cloning		
	28	36	43
pI	7.55	6.93	8.19
Amino acid			
Aspartate	55	80	60
Threonine	57	68	74
Serine	34	32	42
Glutamate	84	67	62
Proline	18	14	26
Glycine	54	36	32
Alanine	76	75	98
Cysteine	14	16	13
Valine	22	18	18
Methionine	9	4	1
Isoleucine	19	18	25
Leucine	52	51	54
Tyrosine	15	20	17
Phenylalanine	11	11	9
Histidine	12	6	12
Lysine	74	69	55
Arginine	7	16	17

* Expressed as residues per 65 000 daltons molecular weight. Tryptophan not measured. Data from Cross and Johnson (1976).

of *T. lewisi* also possess a surface coat, although it is of radically different ultrastructure to that of *T. brucei* and seems limited to only two antigenic types manifest in dividing and adult forms, respectively. There is also some evidence for a surface coat on the blood stages of *T. cruzi*, and evidence for antigenic variation in *Plasmodium* and *Babesia*, although the biochemical basis for this has yet to be studied.

7.5.3 Proteins of microbodies

The peroxisomes (often called microbodies) of mammalian cells are oxidative organelles containing many and various oxidases, including those that act on α-hydroxyacids (glycolate, lactate), purine derivatives (urate, allantoin) and L- and D-amino acids. These oxidases reduce oxygen to hydrogen peroxide, which is metabolised by a haem enzyme called catalase either to O_2 and water or in a peroxidative reaction that oxidises other substrates. This is summarised in figure 7.7. Other enzymes present in peroxisomes include isocitrate dehydrogenase and aminotransferases. The overall function of peroxisomes in mammalian cells is uncertain. Clearly the urate and allantoin oxidases are in-

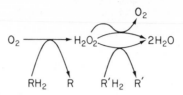

Figure 7.7 Activities of oxidases and catalase in peroxisomes. RH_2, $R'H_2$, reduced substrates.

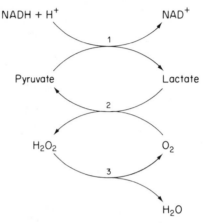

Figure 7.8 Possible role of lactate oxidase in the oxidation of cytoplasmic NADH. Enzymes: 1, lactate dehydrogenase; 2, lactate oxidase; 3, catalase.

volved in purine catabolism. It is possible that lactate oxidase may form a couple with cytoplasmic lactate dehydrogenase that functions in the reoxidation of NADH (figure 7.8).

There are few reports on the microbodies of parasitic protozoa. Catalase is completely absent from blood-stream *Trypanosoma brucei*, although other mammalian trypanosomes (*T. lewisi*) as well as avian, amphibian and insect trypanosomes and *Crithidia* species do possess the enzyme. The hydrogenosomes found in species of *Trichomonas* (section 4.5) are morphologically similar to peroxisomes, but catalase and amino acid oxidases are absent. Catalase is found, however, in the cytoplasm of *T. foetus*, but not in several other species. Catalase is present in the cytoplasm of *Entamoeba invadens*, peroxisomes are absent. D-amino acid oxidases have been reported to be present in *Plasmodium berghei*, buth the localisation has not been studied.

7.5.4 Superoxide dismutase

In many enzyme reactions, oxygen is reduced partially to the superoxide anion (O_2^-), an extremely reactive molecule that can cause irreversible damage to many cell constituents. Superoxide dismutase converts this potentially lethal anion into hydrogen peroxide and molecular oxygen

$$2O_2^- + 2H^+ \longrightarrow H_2O_2 + O_2$$

Only obligatory anaerobic cells do not contain an enzyme that catalyses this reaction, and it has been suggested that aerotolerance is dependent on its presence rather than that of catalase. In the absence of catalase it is presumed that hydrogen peroxide, also potentially lethal, is metabolised by other enzymes possessing peroxidase activity. Mammalian cells contain two superoxidase dismutase enzymes. One is located in the cytoplasm, contains Cu^{2+} and Zn^{2+} and is cyanide sensitive, whereas the other is mitochondrial and is similar to the typical prokaryotic enzyme in containing Mn^{2+} and being insensitive to cyanide. No superoxide dismutase is found in the peroxisomes of mammalian cells.

Trichomonas species also possess two superoxide dismutase enzymes, one localised in the cytoplasm, the other in the hydrogenosomes (a further indication that they are distinct from peroxisomes). Although different from each other, both are cyanide insensitive and are more similar to the prokaryotic enzyme than the cytoplasmic enzyme of mammalian cells. A cytoplasmic superoxide dismutase which is cyanide insensitive and resembles the prokaryotic enzyme has not been reported from any other eukaryotic cell. This enzyme has not been investigated in any other parasitic protozoan.

7.5.5 Polyhistidines in *Plasmodium*

Intraerythrocytic trophozoites of *Plasmodium lophurae* have been found to synthesise a protein of molecular weight 38 000 daltons and containing 73 per cent histidine residues. Within the trophozoites, this protein is contained in cytoplasmic granules, whereas in the merozoites that develop from them, it is localised within the rhoptries and micronemes. This protein has unusual properties and causes the extension and invagination of cell membranes. The exciting possibility is that it is important in the entry of the merozoite into new host red blood cells. The protein is insoluble and inactive at physiological pH, which allows its storage within the malaria cell. Activation is by a change in pH.

7.5.6 Pigment in *Plasmodium*

The intraerythrocytic stage of *Plasmodium* species characteristically produces pigment, which can be seen easily with the light microscope, especially in schizont stage parasites. The composition and function of malaria pigment is unknown. It consists of 1 per cent ferric iron, which is equivalent to a haemin content of 10 per cent, and has a high molecular weight. It has been suggested that it is actively synthesised by the parasite from haemin and other unknown material. However, most recent work with *P. lophurae* has indicated that it is aggregated iron III protoporphyrin IX with contaminating amounts of protein macromolecules bonded to the aggregate by Van der Waals and ionic forces. It is interesting that *Babesia*, which also digests host-cell haemoglobin, does not produce pigment.

7.5.7 Proteins and cyclic AMP

Adenosine-3′,5′-monophosphate (cyclic AMP) is known to be very important in the regulation of growth and development of cells. In mammals, it is believed to be the intracellular chemical messenger between a number of hormones and target enzyme systems within the cell. This activity is mediated by the positive modulation by cyclic AMP of a number of protein kinases. Protein phosphorylation can change enzyme activity, as well as the conformation of membrane, chromatin or ribosomal proteins, with resultant changes in related cellular functions. In bacteria, cyclic AMP mainly functions through enzyme induction and repression, which it affects by activation of the regulatory proteins that bind to promoter genes in the process of induction. In contrast, cyclic AMP does not appear to be involved in enzyme induction and repression in mammalian cells. The level of cyclic AMP is controlled by two enzymes: adenylate cyclase

$$ATP \longrightarrow cyclic\ AMP + PP_i$$

and cyclic AMP phosphodiesterase

$$cyclic\ AMP + H_2O \longrightarrow adenosine\text{-}5'\text{-}phosphate$$

Although little research into this area of metabolism has been carried out with parasitic protozoa, cyclic AMP and the two enzymes have been found in blood-stream trypanosomes. Blood-stream *Trypanosoma gambiense* and epimastigote *T. cruzi* possess protein kinasis that catalyse the phosphorylation of histone and protamine. They are not dependent on cyclic AMP, however, although affected by nucleosides. Blood-stream *T. gambiense* also are permeable to cyclic AMP so it is possible that host hormone levels could indirectly influence their metabolism. The levels of cyclic AMP in blood-stream *T. lewisi* have been shown to rise at the same time as they come under the effect of host cell antibody and undergo morphological change. The significance of these findings to trypanosomes is as yet unknown, but clearly this is an exciting area of research for the future.

7.5.8 Proteins and taxonomy

The structures of proteins have remained remarkably stable during the millions of years over which present-day organisms have evolved. However, alterations have occurred, and the extent by which the proteins of two species differ is a measure of their phylogenetic relationship. Some proteins already have proved very useful to taxonomists—cytochrome *c* in particular. The determination of the primary sequence of a protein is a time consuming process, however, and as yet few proteins from parasitic protozoa have been sequenced. The cytochrome *c* proteins from *Crithidia* species (section 5.2) are exceptions. Proteins that differ in charge, due to different numbers of basic or acidic amino acids, can be distinguished quite simply by electrophoretic methods. The enzymes of species tend to differ and this enables species to be characterised by the mobility of a few of their constituent enzymes on starch gel electrophoretic plates and polyacrylamide gel columns. Many species or subspecies of parasitic protozoa, particularly of the genera *Trypanosoma*, *Plasmodium*, *Eimeria* and *Leishmania*, have been characterised in this way in recent years. Clearly, this method, although not of use in the field, has applications in distinguishing between morphologically similar species. It is useful, also, in the study of the genetics of parasitic protozoa.

7.6 Further Reading

Barker, D. C. (1976). Differentiation of *Entamoebae*: patterns of nucleic acids and ribosomes during encystment and excystation. In *Biochemistry of Parasites and Host–Parasite Relationships* (ed. H. Van den Bossche), Elsevier-North Holland Biomedical Press, Amsterdam, pp. 253–260

Cross, G. A. M. and Johnson, J. G. (1976). Isolation and properties of variant-specific glycoprotein antigens constituting the trypanosome surface coat. In *Biochemistry of Parasites and Host–Parasite Relationships* (ed. H. Van den Bossche), Elsevier-North Holland Biomedical Press, Amsterdam, pp. 413–420

Godfrey, D. G. (1976). Biochemical strain characterisation of trypanosomes. In *New Approaches in American Trypanosomiasis Research*, Pan American Health Organization Scientific Publication No. 318, Washington, pp. 91–96

Hanas, J., Linden, F. and Stuart, K. (1975). Mitochondrial and cytoplasmic ribosomes and their activity in blood and culture form *Trypanosoma brucei*. *J. Cell Biol.*, **65**, 103–111

Kidder, G. W. and Dewey, V. C. (1972). Methionine or folate and phosphoenolpyruvate in the biosynthesis of threonine in *Crithidia fasciculata*. *J. Protozool.*, **19**, 93–98

Kilejian, A. (1976). Studies on a histidine-rich protein from *Plasmodium lophurae* with host and parasite membranes. In *Biochemistry of Parasites and the Host–Parasite Relationships* (ed. H. Van den Bossche), Elsevier-North Holland Biomedical Press, Amsterdam, pp. 441–448

Levy, M. R., Siddiqui, W. A. and Chou, S. C. (1974). Acid protease activity in *Plasmodium falciparum* and *P. knowlesi* and ghosts of their respective host red cells. *Nature*, **247**, 546–549

McLaughlin, J. and Meerovitch, E. (1975). The surface membrane and cytoplasmic membranes of *Entamoeba invadens* (Rodhain 1934)—I. Gross chemical and enzymatic properties. *Comp. Biochem. Physiol.*, **52B**, 477–486

Stibbs, H. H. and Seed, J. R. (1975). Further studies on the metabolism of tryptophan in *Trypanosoma brucei gambiense*. Cofactors, inhibitors and end products. *Experientia*, **31**, 274–277

Walter, R. D. (1975). Inhibition of $3':5'$-cyclic AMP phosphodiesterase from *Trypanosoma gambiense* by deoxyadenosines. *Z. Physiol. Chem.*, **356**, 43–45

8 Lipid Metabolism

8.1 Introduction

Lipids are organic molecules, largely hydrocarbon in structure, that are insoluble in water but soluble in non-polar solvents such as ether, chloroform and methanol. Lipids are essential constituents of all living material, having important roles in cellular structure, energy storage, transport, and metabolic control.

8.2 Structure and Occurrence

Lipids are of two main types—complex and simple. Complex lipids contain fatty acids joined to a backbone, and according to the structure of their backbone are classified as acylglycerols, phosphoglycerides, sphingolipids or waxes. Simple lipids, which contain no fatty acids, include the terpenes, steroids and prostaglandins. Both types of lipid are present in most cells, but the relative abundance and detailed structure varies. The total lipid content and the relative abundance of the different lipids reported for the few parasitic protozoa investigated are shown in table 8.1. Generally the free fatty acid and acylglycerol content is low whereas the phosphoglyceride and sphingolipid content is high, that is structural rather than storage lipids are predominant.

8.2.1 Fatty acids

Fatty acids occur in all complex lipids, although, normally, only very low concentrations of free fatty acids are present in cells. The basic structure of fatty acids is shown in figure 8.1. More than 100 different fatty acids are known, but some are found more frequently than others. Palmitic ($16:0$) and stearic ($18:0$) are common saturated fatty acids, oleic ($18:1^{\Delta 9}$), linoleic ($18:2^{\Delta 9,12}$) and arachidonic ($20:4^{\Delta 5,8,11,14}$) are abundant unsaturated fatty acids. Linoleic acid is the most common fatty acid in mammals, accounting for 10–20 per cent of those present.

The occurrence of different fatty acids in parasitic protozoa has been reported in a few cases (table 8.2.). However, it has been shown with several parasitic protozoa, especially *Entamoeba* species, that the fatty acid content is a reflection of the fatty acids available in the environment, and in different con-

Table 8.1 Lipid content of some parasitic protozoa

	Lipid content of cell (% of dry weight)	Percentage of total lipids		
		Phosphoglycerides*	Acylglycerols Free fatty acids	Simple lipids
Trypanosoma rhodesiense B	11	79	4	17
Trypanosoma rhodesiense C	17	73	3	21
Trypanosoma brucei B	16			
Trypanosoma lewisi B	12	79	4	17
Trypanosoma lewisi C	11	72	11	17
Trypanosoma cruzi C	20			
Eimeria acervulina O	14			
Plasmodium knowlesi B	29			

* Including sphingolipids; B, blood stages; C, culture stages; O, oocysts

Saturated CH_3—$(CH_2)_n$—COOH

Monounsaturated CH_3—$(CH_2)_m$—CH$=$CH—$(CH_2)_l$—COOH

Figure 8.1 Structure of fatty acids. Examples: palmitic acid, 16 carbons saturated (16:0); oleic acid, 18 carbons and one double bond at carbons 9 and 10 ($18:1^{\Delta 9}$).

ditions it alters markedly. Of interest is the discovery that species of *Crithidia* and *Leishmania* contain cyclopropane (9,10-methyleneoctadecanoic acid) in some phosphoglycerides, although it has not been found yet in other Kinetoplastida. Previously this type of acid had been found only in bacteria. Nevertheless there is clearly no radical difference between the fatty acid composition of parasitic protozoa and mammalian cells.

8.2.2 Acylglycerols

Acylglycerols are fatty acid esters of glycerol. The most common acylglycerols contain three fatty acids, and are called triacylglycerols. A large number of these exist, each with different constituent fatty acids. They are the most abundant group of lipids and the major component of storage lipid in mammalian cells such as adipose cells.

The occurrence of triacylglycerols in parasitic protozoa is not well documented. Low levels are present in blood-stream trypanosomes, and somewhat higher levels in the culture forms. Erythrocytes infected with *Plasmodium knowlesi* contain triacylglycerols, diacylglycerols and free fatty acids at concentrations higher than those present in normal red blood cells. The significance of this is not known. *Eimeria* oocysts possess lipid stores, which

Table 8.2 Fatty acids identified in some parasitic protozoa

	Trypanosoma rhodesiense B [b,d]	Trypanosoma lewisi B [c]	Crithidia fasciculata [b]	Leishmania donovani C [c]	Leishmania tropica major C [c]	Leishmania enriettii C [c]	Entamoeba invadens [c]	Plasmodium lophurae B [b]	Plasmodium berghei B [b]	Rat plasma phosphoglycerides	Rat plasma triacylglycerols
10:0								(+)			
11:0								(+)			
12:0			0.6	0.2	0.2	(+)		+	+		
12:1								(+)	+		
13:0			0.2	0.1	0.1	0.4		(+)	(+)		
14:0	+	2	5	3	3	5	7	+	+		
14:1								+	+		
14:2								(+)	(+)		
15:0 iso			2	2	2	3					
16:0	19	16	4	8	8	9	8	+	+	26	32
16:1	(+)	2	1				7	+	+		
16:2								+	(+)		
17:0			2	0.2	0.1	−		+	+		
17:0 iso			1								
18:0	13	22	11	14	10	7	0.6	+	+	15	0.9
18:1	5	26	26	25	32	24	56	+	+	8	30
18:2	38	26	10	30	28	23	6	+	+	25	22
18:3	0.4	(+)	17	15	16	20		+	+	(+)	2
20 poly [a]	6	(+)	1					+	+	15	3
22 poly [a]	16	(+)	11							9	9
24							8				

B, blood stages; C, culture forms; a, all acids with two or more double bonds; b, fatty acids in phosphoglycerides only; c, fatty acids in total lipids; d, blood-stream long slender form; +, acid present; (+) acid present in traces; −, acid absent; all figures are percentages of total fatty acids

are important sources of energy for sporulation, and probably consist mainly of triacylglycerols.

8.2.3 Phosphoglycerides

Phosphoglycerides (figure 8.2) are lipids characteristically present in abundance in cell membranes (which contain approximately 40 per cent lipid and 60 per cent protein) but only small amounts are found elsewhere in the cell. They possess both polar and non-polar ends, and it is this amphipathic nature that is responsible for much of their importance to the cell. Various polar head groups are found, the most common are choline and ethanolamine. These two phosphoglyceride types are the major lipid components of most animal cell membranes. Other important phosphoglycerides are phosphatidylserine and phosphatidylinositol. Cardiolipin is a phosphoglyceride characteristically abundant in the cell membranes of bacteria, and is found also in large amounts in the inner membrane of mitochondria. Each type of phosphoglyceride can exist in many different chemical species differing in their fatty acid content, although

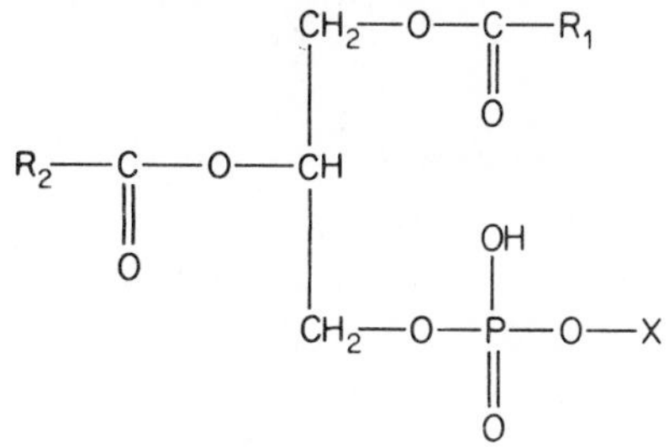

Figure 8.2 Structure of phosphoglycerides. R_1, R_2, hydrocarbon chains of fatty acids; X, polar head group. Examples: phosphatidic acid, X=H; phosphatidylcholine, X=choline; phosphatidylethanolamine, X=ethanolamine.

normally there is one saturated and one unsaturated fatty acid. Another class of phosphoglycerides are the plasmalogens, in which one of the hydrocarbon tails is a long aliphatic chain in *cis*-α,β-unsaturated ether linkage at the 1 position of the glycerol. In mammals these are especially abundant in the membranes of muscle and nerve cells.

It is very probable that all parasitic protozoa possess phosphoglycerides, but the actual types present have been investigated in only a few cases (table 8.3). Phosphatidylcholine is the predominant type present in most trypanosomes; phosphatidylethanolamine is also common. Phosphatidylserine and relatively high levels of phosphatidylinositol have been reported in *Trypanosoma rhodesiense* blood-stream and culture forms. *Crithidia fasciculata*, although containing phosphatidylcholine and phosphatidylethanolamine, possess no phosphatidylserine. However, phosphatidylinositol is found at much higher levels than in mammalian cells. Triphosphoinositide and diphosphoinositide are also present, making up greater than 1 per cent of the total phosphoglycerides. The turnover of these polyphosphoinositides is very rapid and involves polyphosphoinositide phosphodiesterase and polyphosphoinositide phosphatase which are broadly similar to the mammalian brain and kidney enzymes. These polyphospoinositides were initially found in mammalian nervous tissue, but more recently have been identified in other tissues and yeast. Their role is still a mystery.

Less is known of the phosphoglycerides of other parasitic protozoa. Intraerythrocytic stages of *Plasmodium knowlesi* contain phosphatidylcholine, phosphatidylethanolamine and phosphatidylinositol but no phosphatidylserine. The membranes of the parasite are slightly richer in

Table 8.3 Phosphoglycerides identified in some parasitic protozoa

	Phosphatidylcholine	Phosphatidylethanolamine	Phosphatidylserine	Inositol phosphatides	Phosphatidic acid	Cardiolipin	Plasmalogens	Sphingolipids
Crithidia fasciculata	+	+	−	+		+		+
Trypanosoma rhodesiense B	+	+	+	+		−	+*	+
Trypanosoma rhodesiense C	+	+	+	+		+		+
Trypanosoma vivax B								+
Trypanosoma cruzi C	+	+		+			+	
Trichomonas vaginalis						+		
Entamoeba invadens	+	+	+	+	+			+
Plasmodium knowlesi B	+	+	−	+				+

B, blood stages; C, culture forms; *, short stumpy forms only; +, present; −, absent

phosphoglycerides than those of the host red blood cell. The malaria parasite hydrolyses the host cell phosphoglycerides (which include phosphatidylserine). This activity may be concerned with the increase in host cell permeability induced by the parasite (section 4.8) and may account also for the increased fragility of the infected erythrocytes compared with normal red blood cells. Furthermore, it may be one cause of the haemolysis found with malaria infections, for the hydrolytic products (lysophosphatidylethanolamine and lysophosphatidylcholine) have haemolytic properties and may damage uninfected erythrocytes.

Entamoeba invadens contains at least eleven phosphoglycerides and sphingolipids. The most abundant phosphoglyceride is phosphatidylethanolamine. Phosphatidylcholine is the other major type and small amounts of phosphatidylserine, phosphatidylinositol and phosphatidic acid are also present. These last two have relatively short half lives, of the order of 10 h, which may be correlated with the turnover of membranes in this phagocyte, as has been shown in other such cells. Other phosphoglycerides turn over either slowly or not at all in the case of phosphatidylserine and phosphatidylcholine. Little is known of the phosphoglycerides of *Trichomonas*; however, cardiolipin is present in the hydrogenosomes.

Plasmalogens occur in abundance in many strictly anaerobic organisms including rumen protozoa. They have been reported to occur also in *Leishmania*, *Trypanosoma cruzi* and short stumpy blood-stream *T. rhodesiense* but to be absent from the long slender forms and from *Crithidia fasciculata*.

8.2.4 Sphingolipids

Sphingolipids are important membrane components, present in especially large amounts in brain and nervous tissue in mammals. Structurally they are composed of a backbone of sphingosine or a related base, one molecule of fatty acid and a polar head group, which in some sphingolipids is very large and complex. As with phosphoglycerides, many different sphingolipids exist.

Sphingolipids have been demonstrated in several trypanosomatids (table 8.3). *Trypanosoma vivax* is reported to possess over 50 per cent of its lipid phosphorous as spingomyelin. The sphingolipids of *Crithidia fasciculata* have been found to contain the novel base 19-methyl-C_{20}-phytosphingosine. *Entamoeba invadens* possesses relatively high concentrations of sphingolipids, approximately 30 per cent of the concentration of phosphoglycerides. Four types have been identified, sphingomyelin (a minor component), ceramide phosphorylethanolamine, ceramide aminoethylphosphonate and the previously unrecorded ceramide phosphonylinositol. Phosphonolipids are highly resistant to hydrolysis. Their abundance in the surface membrane of *Entamoeba* may be related both to the surface-active lysosomes present (sections 7.2 and 8.3) and the tissue invasion phase of the parasites life cycle. The sphingolipid content of *E. invadens* is dependent on the environment. It is adjusted to maintain membrane integrity in conditions of different fatty acid availability. Sphingolipids have also been found in species of *Plasmodium*.

8.2.5 Waxes

No waxes have been found in parasitic protozoa.

8.2.6 Simple lipids

Simple lipids are less abundant than complex lipids; nevertheless they include many compounds with important biological activities in mammals, such as vitamins and hormones. Simple lipids are structurally all related, being derived from a five carbon building block. Two classes of simple lipids are recognised, the terpenes and the steroids. The only report of terpenes in parasitic protozoa is the possible existence of squalene—an important precursor of cholesterol—in *Crithidia fasciculata*.

In mammals, steroids include some very important hormones as well as the bile acids. A large subgroup of steroids are the sterols which include cholesterol (figure 8.3). This is the most abundant steroid in animal tissues, and occurs in the plasma membrane of many cells and in the lipoproteins of blood plasma. It is also the precursor of many hormones and bile salts. Sterols can exist in two forms, as free alcohols and as long-chain fatty acid esters of the hydroxyl group at carbon 3. Sterols are not present in bacteria. The main sterols of plants are the phytosterols, whereas fungi and yeast contain the mycosterols, which include ergosterol.

Cholesterol is the predominant sterol, and steroid, in many parasitic protozoa, including *Trichomonas*, *Plasmodium* and *Entamoeba*. The surface membrane of *Entamoeba* is particularly rich in it. Cholesterol is the major sterol of blood-stream trypanosomes, and accumulates in the short stumpy forms of *Trypanosoma rhodesiense*. The significance of this accumulation and the various other differences in lipid content between the short stumpy and long slender blood-stream forms is unclear. Ergosterol is the most abundant sterol in the culture forms of many trypanosomatids, including *T. lewisi*, *T. cruzi* and *T.*

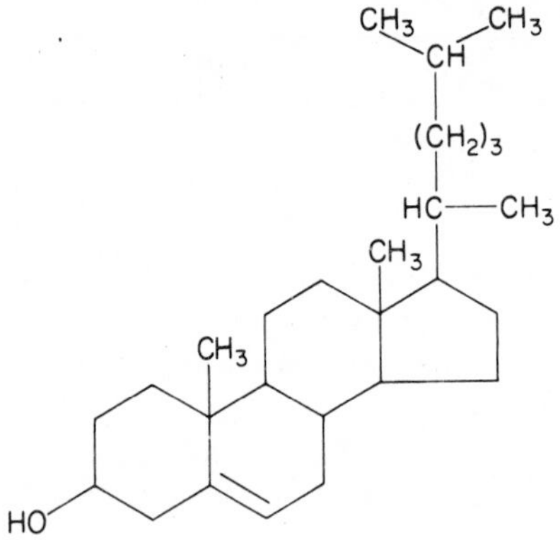

Figure 8.3 Cholesterol.

rhodesiense, although cholesterol also is present in many cases. The sterol content of many parasitic protozoa is a reflection of the exogenously available sterols, for many species are unable to synthesise sterols *de novo* which, consequently, are an essential dietary requirement. As an example, *T. cruzi* growing in a cholesterol-deficient medium contains ergosterol, whereas growth in the presence of abundant cholesterol results in this becoming the most common sterol in the organism. The sterols identified in parasitic protozoa are shown in table 8.4. In most species sterols are predominantly free; however, sterol esters do occur. The fatty acids involved are very diverse, and not necessarily similar to those present in the phosphoglycerides. For instance, the C_{20}–C_{24} acids make up approximately 40 per cent of the total acids of the sterol esters in *Plasmodium lophurae* and *P. berghei*, but only 5 per cent of the phosphoglyceride fatty acids.

Prostaglandins are a group of lipids, derived from fatty acids, which are important in mammalian cell regulation. None has been found in parasitic protozoa.

Acetylcholine is a very important messenger in the nervous system of mammals. It has been found in *T. rhodesiense* but is absent from *P. gallinaceum*. Its role is unclear.

Table 8.4 Sterols identified in some parasitic protozoa

	Cholesterol	Ergosterol	Spinasterol	Dihydroergosterol	Dehydroporiferasterol	Dehydroclionasterol	Uncharacterised sterols
Crithidia fasciculata	+	+		+			
Crithidia oncopelti		+	+				
Trypanosoma rhodesiense B	+						
Trypanosoma rhodesiense C	+	+					+
Trypanosoma cruzi B	+						+
Trypanosoma cruzi C	+	+		+	+	+	
Trypanosoma lewisi B	+						+
Trypanosoma lewisi C	+	+					+
Trichomonas vaginalis	+						
Entamoeba histolytica	+						
Plasmodium knowlesi B	+	+					

B, blood stages; C, culture forms; +, present.

8.3 Utilisation of Exogenous Lipids

Exogenous lipids are potentially useful to a cell either directly or as sources of energy or lipid precursors. Mammals are capable of synthesising most of the lipids they require. Exceptions are the two polyunsaturated fatty acids linoleic ($18:2^{\Delta 9,12}$) and linolenic ($18:3^{\Delta 9,13,15}$) acids, which are essential dietary requirements. They are used not only *per se* for incorporation into lipids but also as precursors in the synthesis of prostaglandins. Mammals, however, do utilise other lipids, if available, both for energy and in synthesis. Lipids in general but triacylglycerols in particular have a high energy content that is largely released, in a utilisable form, on hydrolysis. This hydrolysis is catalysed by lipases, which, in mammals, are under hormonal control. Fatty acids provide approximately 40 per cent of the total energy requirement of a man on a normal diet.

Most parasitic protozoa apparently require at least some preformed lipids; however, *Crithidia* and *Leishmania tarentolae* can be cultured in the absence of lipid. In most culture media this requirement is satisfied by the addition of unfractionated serum. With some media it has been possible to replace serum with defined lipids, thus indicating those essential to the growth of the organism concerned. With the exception of most trypanosomatids, which can synthesise sterol *de novo*, and *L. tarentolae*, all parasitic protozoa investigated have been shown to have obligate dietary requirements for both a sterol (normally cholesterol) and at least one fatty acid. The serum in media for *Trichomonas* can be replaced by cholesterol, oleic and palmitic acids, and that in media for *Entamoeba* by an unsaturated fatty acid (oleic supports best growth and is the most abundant fatty acid in the phosphoglycerides) and cholesterol. *Plasmodium knowlesi* will develop in a medium in which the only lipids are stearic acid and cholesterol; however, serum supports better growth. *Trypanosoma cruzi* culture epimastigotes apparently will grow in a medium containing only stearic acid.

Other lipids than those essential to the growth of the parasitic protozoa are absorbed and utilised if available. Blood-stream trypanosomes will absorb fat droplets as well as saturated and unsaturated fatty acids. Exogenous fatty acids may be an important source of energy to *T. cruzi* blood-stream stages, while *Plasmodium* species also take up saturated and unsaturated fatty acids. Although some Kinetoplastida can synthesise sterols *de novo*, they will salvage them if available.

The mechanisms of uptake of lipids into parasitic protozoa are unknown. Presumably, once absorbed, they are hydrolysed by lipases; intracellular particle-bound lipases have been found in trypanosomes. Secretion of lipases into the environment has not been demonstrated with any parasitic protozoan. *Entamoeba histolytica* and other *Entamoeba* species, however, do contain surface active lysosomes that are capable of lysing cellular material which comes into contact with them. They do not excrete enzymes into the medium, but possibly contain a membrane-bound lipase which disrupts the membrane of cells in contact with it. The intraerythrocytic stages of *P. lophurae* possess phospholipase A.

8.4 Biosynthesis of Lipids

The biosynthesis of lipids is an important process in most cells and the mechanisms involved in mammalian cells have been characterised in detail. Lipid biosynthesis can be considered in three parts: biosynthesis of fatty acids; their incorporation into complex lipids; the biosynthesis of simple lipids.

8.4.1 Fatty acids

Mammalian cells, especially of liver, adipose tissue and mammary gland, synthesise fatty acids from acetyl CoA by a series of reactions that are catalysed by a multienzyme complex, called fatty acid synthetase, which is present in the cytoplasm. This synthesis involves the successive addition of two-carbon units to the growing fatty acid chain. The process is quite distinct in mechanism and localisation from the catabolism of fatty acids (section 8.5). The normal end product of the fatty acid synthetase system is palmitic acid, and this is the precursor of other long chain saturated and unsaturated fatty acids. Elongation of fatty acids is catalysed by two distinct enzyme systems. One, in the mitochondria, uses acetyl CoA as the source of two-carbon units whereas the other, in the endoplasmic reticulum, uses malonyl CoA. Unsaturated fatty acids are produced by a monooxygenase system present in the endoplasmic reticulum of mammalian cells. Bacteria unsaturate fatty acids by a different, anaerobic mechanism.

No parasitic protozoan has been demonstrated to carry out the *de novo* synthesis of fatty acids, although such a pathway probably exists in some Kinetoplastida. Chain elongation, however, probably occurs in all species as is shown, for example, by the incorporation of acetate into fatty acids by *Trichomonas foetus*, malaria parasites and various Kinetoplastida. In *Trypanosoma* and *Crithidia* but not apparently in *Leishmania*, the acetate units used in fatty acid synthesis (where it occurs) and chain elongation are derived mainly from threonine. The probable pathway involved is shown in figure 8.4. This

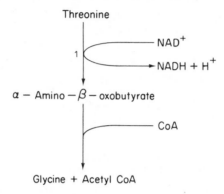

Figure 8.4 Possible mechanism for the conversion of threonine to acetate in trypanosomatids. Enzyme: 1, threonine dehydrogenase

conversion is strongly inhibited by tetraethylthiuram disulphide which is also a potent growth inhibitor of *Trypanosoma brucei* culture forms (50 per cent inhibition at 7×10^{-7} M). This indicates the importance of threonine metabolism to these organisms. Desaturation of fatty acids has been demonstrated also in trypanosomatids. Little is known of the mechanisms of either chain elongation or desaturation, although biopterin is involved as a cofactor in the desaturation of fatty acids by *Crithidia fasciculata*. Study of the synthesis of the unusual fatty acids of *C. fasciculata* indicates that the iso acids (table 8.3) are formed by chain elongation of isovaleryl CoA which in turn is derived from leucine. The C_{19} cyclopropane acid is synthesised from oleic acid, which serves as an acceptor for the methyl group of methionine.

8.4.2 Triacylglycerols

In mammals, triacylglycerols are synthesised by the successive acylation of L-glycerol-3-phosphate by fatty acyl CoA to give phosphatidic acid (figure 8.2). This undergoes hydrolysis to diacylglycerol which is further acylated by another fatty acyl CoA to triacylglycerol. Nothing is known of the mechanism of synthesis of triacylglycerols by parasitic protozoa, except that L-glycerol-3-phosphate is readily synthesised by most species.

8.4.3 Phosphoglycerides

In mammalian cells, phosphoglycerides are synthesised from phosphatidic acid (figure 8.2) in reactions catalysed by enzymes bound to the endoplasmic reticulum. Cytidine nucleotides serve as carriers in these reactions.

Many parasitic protozoa synthesise phosphoglycerides quite readily, including species of *Trypanosoma*, *Crithidia*, *Plasmodium* and *Entamoeba*. Interestingly, the blood-stream form of *T. cruzi*, which does not divide, incorporates fatty acids into phosphoglycerides at a very low rate or not at all, whereas culture forms carry out the incorporation readily. The pathways present in *C. fasciculata* are similar to those of mammalian cells. Phosphatidylcholine is synthesised both by methylation of phosphatidylethanolamine, a process which occurs even when preformed choline is available, and by the phosphocholine transferase pathway which is the dominant mechanism in mammalian cells. Phosphatidylethanolamine is produced both by a phosphoethanolamine transferase pathway, characteristic of mammalian cells, and probably by decarboxylation of phosphatidylserine. A very active decarboxylase is present, and this could explain the apparent paradox that serine is a good precursor of phosphoglycerides in *C. fasciculata* and yet the cells contain no phosphatidylserine.

The fatty acid content of the phosphoglycerides of *Entamoeba invadens* reflects the availability of exogenous fatty acids, especially unsaturated ones. Similarly the level of sterols in the organism reflects the concentration in the medium. As the exogenous concentration of cholesterol is increased, so the ratio of cholesterol to phosphoglycerides in the organism rises. However, it is essential to retain the correct fluidity of the cell membranes, and it has been suggested that this is achieved by adjusting the ratio of unsaturated to saturated fatty acids in the phosphoglycerides. This is possibly controlled at the level of the enzymes catalysing the incorporation of fatty acids into lipids.

The mechanism of incorporation of glycerol into plasmalogens by anaerobic rumen protozoa is the same as that operative in anaerobic bacteria but different from that in mammalian cells. If similar pathways are present in anaerobic parasitic protozoa, they are potential targets for chemotherapeutic attack.

8.4.4 Sterols

The synthesis of sterol from acetyl CoA by mammalian cells is a complex process involving some 25 reaction steps, which will not be discussed here.

Most parasitic protozoa cannot synthesise sterols and obtain their requirement from the environment. For this reason the sterol content of many species is controlled by the availability of exogenous sterols. Some metabolism of sterols probably occurs, however. This is indicated by the finding that several sterols, besides cholesterol, and some sterol derivatives can support the growth of trichomonads. These organisms contain a dehydrogenase that specifically interconverts 17-keto and 17-hydroxyl groups of certain C_{18} and C_{19} steroids.

Among the parasitic protozoa, *de novo* synthesis of sterols is limited to the Kinetoplastida. *Crithidia fasciculata* incorporates acetate, mevalonate and methionine into ergosterol, and *T. lewisi* blood-stream and culture forms and *T. rhodesiense* culture forms (but not blood-stream forms) synthesise sterol from acetate and methionine. The pathways of *de novo* synthesis are unknown. Thus trypanosomatids synthesise some of the sterols they contain. However, cholesterol probably is not synthesised *de novo*, and when present is of dietary origin.

The possibility of a correlation between the absence of sterol biosynthetic ability in blood-stream *T. rhodesiense* and the lack of functional mitochondria is supported by the finding that other sterols and a biosynthetic pathway are present in stages and species possessing cristate mitochondria.

8.5 Catabolism of Lipids

Mammals hydrolyse complex lipids in reactions catalysed by lipases. The fatty acids produced are important sources of energy for mammalian cells, and are oxidised in all mammalian tissues except the brain. Catabolism of fatty acids, which occurs in the mitochondria, involves several enzymes and the whole process is called β-oxidation. The immediate products are acetyl CoA and reduced coenzymes (NADH and $FADH_2$) all of which can be oxidised further to produce energy. Complete oxidation of a C_{12} saturated fatty acid yields 95 molecules of ATP.

Little is known of the catabolism of lipids in parasitic protozoa. Our present knowledge indicates that generally lipids are less important as an energy source in parasitic protozoa than in mammals. Fatty acids may be an important energy source to blood-stream *T. cruzi*, which will catabolise several fatty acids to CO_2 by an uncharacterised pathway, but are not to culture forms. It has been suggested that this utilisation of fatty acids may be a factor in determining the distribution of *T. cruzi* amastigotes. They are mainly found in striated and cardiac muscle which possess similar fat metabolism to *T. cruzi* blood-stream stages. These tissues also contain high levels of carnitine which may play an important part both in the uptake of fatty acids by *T. cruzi* and also in the differentiation of the parasites. *T. rhodesiense* and *T. lewisi* blood-stream forms both oxidise fatty acids but to a very small extent. The culture forms oxidise slightly greater but still small amounts. The mechanisms of catabolism are unknown. Evidence suggests that *C. fasciculata* catabolises fatty acids by both β-oxidation and α-oxidation, the latter pathway being characteristic of germinating plant seeds. However, the details of neither pathway have been investigated. The mechanism by which *Eimeria* oocysts catabolise lipids during sporulation has not been studied.

Fatty acids are excreted by some protozoa, including *Leishmania*. The fatty acids involved differ with species and it has been suggested that these differences could be a useful diagnostic tool.

8.6 General Conclusions

All the major types of structural lipids found in mammalian cells have also been found in some parasitic protozoa. The lipids from the two sources are basically similar in structure, although there are detailed differences. Storage lipids and lipids with important biological activities, such as hormones, are uncommon in parasitic protozoa. The ability to synthesise fatty acids and sterols is limited to some members of the Kinetoplastida. Most parasitic protozoa require them preformed. However, interconversions of fatty acids and sterols and incorporation of fatty acids into complex lipids are carried out. Little detail is known of the pathways involved in lipid metabolism.

8.7 Further Reading

Dewey, V. C. (1967). Lipid composition, nutrition and metabolism. In *Chemical Zoology, vol. 1, The Protozoa* (ed. G. W. Kidder), Academic Press, New York, pp. 161–274

McLaughlin, J. and Meerovitch, E. (1975). The surface membrane and cytoplasmic membranes of *Entamoeba invadens* (Rodhain 1934)—II. Polypeptide and phospholipid composition. *Comp. Biochem. Physiol.*, **52B**, 487–498

Palmer, F. B. St C. (1974). Biosynthesis of choline and ethanolamine phospholipids in *Crithidia fasciculata. J. Protozool.*, **21**, 160–163

Van Vliet, H. H. D. M., Op den Kamp, J. A. F. and Van Deenen, L. L. M. (1975). Phospholipids of *Entamoeba invadens. Archs Biochem. Biophys.*, **171**, 55–64

Venkatesan, S. and Ormerod, W. E. (1976). Lipid content of the slender and stumpy forms of *Trypanosoma brucei rhodesiense*: a comparative study. *Comp. Biochem. Physiol.*, **53B**, 481–487

Wood, D. E. and Schiller, E. L. (1975). *Trypanosoma cruzi*. Comparative fatty acid metabolism of the epimastigotes and trypomastigotes *in vitro. Expl Parasit.*, **38**, 202–207

9 Biochemical Mechanisms of Drug Action

9.1 Introduction

Extensive study of the biochemical modes of action of antimicrobial drugs over many years, has led to the identification of six key areas of metabolism as targets for drug action: energy metabolism; membrane function; cofactor synthesis; nucleic acid synthesis; protein synthesis; cell wall synthesis (table 9.1). Note the absence of lipid metabolism and carbohydrate synthesis from this list. As explained in chapter 1, there is no real equivalent to the bacterial and fungal cell walls in parasitic protozoa. With this exception, it does appear, although from much less extensive studies, that the same key areas are also the targets for antiprotozoan drugs.

Table 9.1 Mechanisms of action of antimicrobial drugs with examples from antibacterial chemotherapy

Mechanism of action	Antibacterial example
Energy metabolism	—
Membrane function	Polymyxin B
Cofactor synthesis	Sulphonamides, trimethoprim
Nucleic acid synthesis	Rifampicin, nalidixic acid
Protein synthesis	Chloramphenicol, streptomycin, tetracyclines, lincomycins
Wall synthesis	Penicillins, cephalosporins

The proper establishment of the primary mode of action of a drug requires a systematic study of its effect on the various metabolic processes of the cell at the lowest concentrations that inhibit growth. Only when such a survey is complete can it be concluded confidently that a particular pathway or reaction is most sensitive to inhibition by a drug and is therefore the primary target of that drug. Systematic studies of this type have been carried out with a number of antibacterial drugs, but only rarely with antiprotozoan drugs. Thus, the assignments made in this chapter of each group of drugs to a particular mode of action are tentative in many cases. There are some drugs for which there is in fact still no clue as to their mode of action.

Investigations of antibacterial drugs have shown that establishment of the biochemical mode of action of a drug often provides an explanation also of why the pathogen is killed but the host is unharmed, that is the mechanism of selective toxicity. This has in many cases also proved to be true with antiprotozoan drugs. Five basic mechanisms have been found to underlie the selectivity of antimicrobial drugs: drug only permeates to, or is concentrated in, the pathogen; drug is metabolised to the active species only in the pathogen; drug attacks a target present only in the pathogen; drug discriminates between isofunctional targets in the pathogen and the host; drug blocks a target that is of greater importance to the pathogen than to the host (table 9.2). There are examples of antiprotozoan drugs in each of these categories.

Table 9.2 Mechanisms of selective toxicity of antimicrobial drugs with examples from antibacterial chemotherapy

Mechanism of selective toxicity	Antibacterial example
Selective permeability	Tetracyclines
Drug activation *in situ*	Nitrofurantoin
Unique target	Penicillins, cephalosporins, sulphonamides
Target discrimination	Trimethoprim, rifampicin, nalidixic acid, chloramphenicol, streptomycin, lincomycins
Target more important	—

9.2 Drugs Interfering with Energy Metabolism

A number of antiprotozoan drugs are thought to exert their antimicrobial action by interference with energy metabolism. These include the arsenicals, antimonials, sulphated naphthylamines, robenidines, quinolones, naphthoquinones and 8-aminoquinolines.

Of these, arsenicals such as tryparsamide and Mel B (figure 9.1) used in the

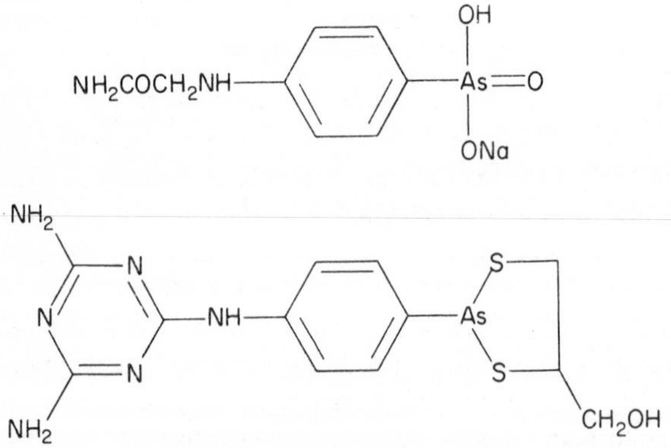

Figure 9.1 Chemical structure of tryparsamide and Mel B.

treatment of African sleeping sickness, have been most investigated. Note that pentavalent forms must be reduced to the trivalent derivative before they are active. It has long been considered that trivalent arsenicals exert their various biological activities by reacting with the sulphydryl groups of proteins. Since many enzymes have SH-groups at their active sites, such a reaction usually has marked inhibitory effects on their activity. At one time or another, all of the glycolytic kinases of trypanosomes have been suggested as the primary target. Most recent work suggests that pyruvate kinase is the most likely site of inhibition. Since the long slender blood-stream forms of African trypanosomes rely exclusively on glycolysis for energy (section 4.3), it is not surprising that the arsenicals thus exert a rapid trypanocidal action on these organisms. The selectivity of the arsenicals appears to rest in part on their greater permeability into trypanosomes, in part on the greater sensitivity of the trypanosomal pyruvate kinase to inhibition and in part on the greater dependence of the trypanosome on glycolysis for ATP synthesis.

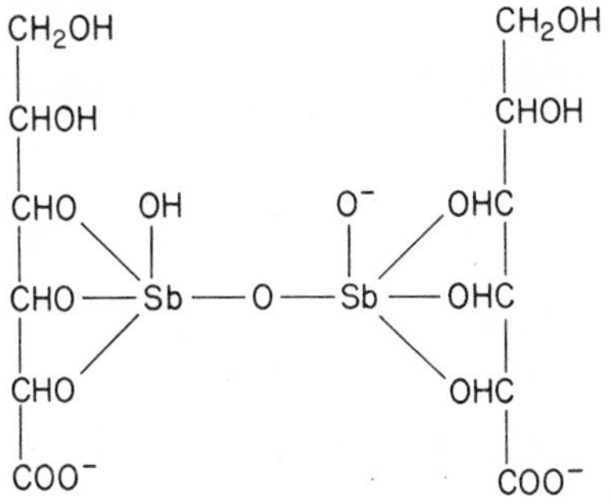

Figure 9.2 Chemical structure of pentostam.

The antileishmanial antimonials such as pentostam (figure 9.2) have not been studied in great detail by biochemists though it is generally considered that they, as with the arsenicals, must be reduced to the trivalent form and that they interact with glycolytic sulphydryl enzymes and thus block energy metabolism. Work on the mode of action of antischistosomal antimonials suggests that the most sensitive glycolytic enzyme is phosphofructokinase. It is likely but not proven that the antileishmanial antimonials act at the same point and that the mechanism of selective toxicity is similar to that of the arsenicals.

The mode of action of the African trypanocide, suramin (a sulphated naphthylamine, figure 9.3), is still obscure. It is known to inhibit many enzymes in cell-free systems at about 10^{-3} M. Now, however, it has been reported that the purified L-α-glycerophosphate oxidase of African trypanosomes, which is concerned with the oxidation of NADH produced during glycolysis in the blood stages (section 4.3), is markedly sensitive to inhibition by suramin ($K_i = 4.1$ μM). It was further shown in studies with a series of analogues of suramin that activity against the purified enzyme paralleled that against infections in mice. It is on the basis of this evidence that the drug is included in this section. However, it must be realised that first, suramin does not reduce oxygen consumption by washed cell suspensions of trypanosomes (this could be a result of slow drug penetration), secondly, there is evidence that neither SHAM (see section 4.3), an inhibitor of the L-α-glycerophosphate oxidase which actually blocks oxygen consumption by washed cell suspensions of trypanosomes, nor anaerobic conditions are trypanocidal and thirdly, at least two other enzymes,

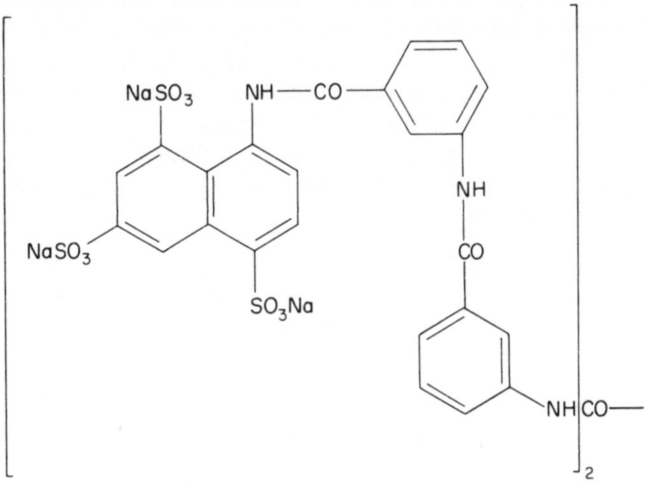

Figure 9.3 Chemical structure of the half molecule of suramin.

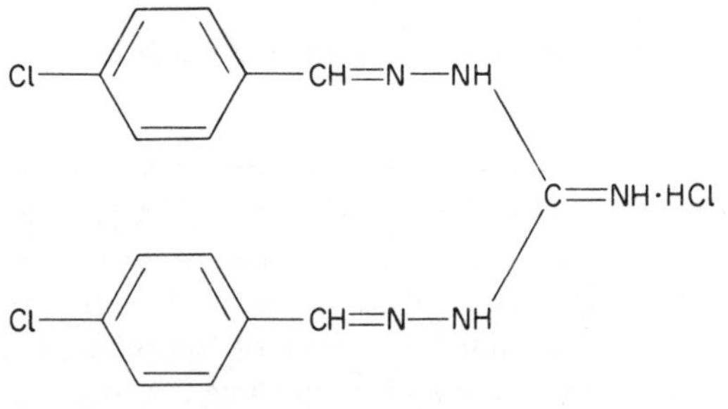

Figure 9.4 Chemical structure of robenidine.

dihydrofolate reductase and thymidylate synthetase (section 6.4) are inhibited strongly by the drug at similar concentrations. The present conclusions about the mode of action of suramin may therefore need later revision. If it is indeed a specific inhibitor of L-α-glycerophosphate oxidase, its mechanism of selective toxicity is explained since there is no mammalian equivalent of this oxidase system.

The anticoccidial drug, robenidine (figure 9.4), has been shown to inhibit

respiratory chain phosphorylation and ATPase activity in rat liver mitochondria. The relevance of this finding to its anticoccidial activity has yet to be established, but for the moment it too is included in this section of inhibitors of
energy metabolism. Since its mode of action is not properly understood, no
speculation can be offered as to its mechanism of selective toxicity.

The antibacterial quinolones resemble drugs such as nalidixic acid which are
thought to have a specific action on DNA synthesis. The anticoccidial
quinolones, including buquinolate, decoquinate and methylbenzoquate (figure
9.5) are, however, structurally distinct and recent evidence suggests that these

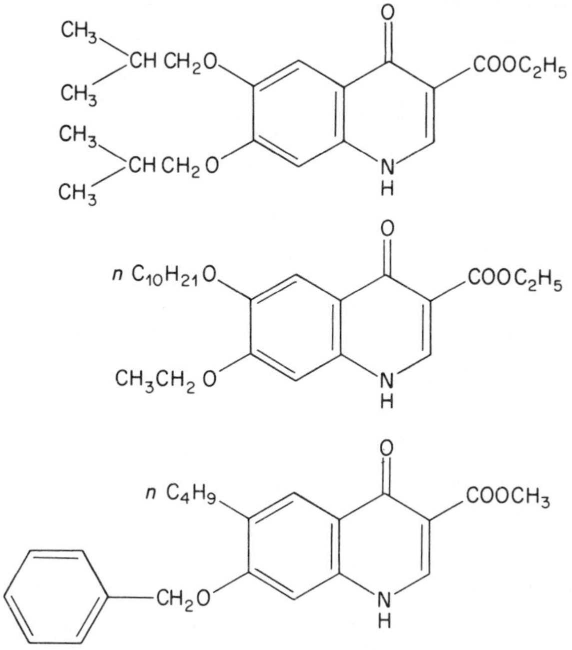

Figure 9.5 Chemical structures of anticoccidial quinolones.

drugs block the respiratory chain of coccidia at a point near to cytochrome b.
Thus the respiration of mitochondria isolated from *Eimeria* oocysts was inhibited 50 per cent at a methylbenzoquate concentration of 10^{-8} M using either
succinate or malate and pyruvate as substrate. Avian mitochondria were un-
affected by much higher drug concentrations, thus providing an explanation of
the selective action of these compounds. However, it should be stressed that it
is still not entirely clear as to how such a mechanism provides an explanation
for the antitrophozoite action of these compounds since it is suspected that
energy metabolism beyond glycolysis does not occur until the schizont stage
(section 4.7). Possible explanations here are either that the respiratory chain is
involved in other metabolic reactions (for example, the dihydroorotate
dehydrogenase reaction in UMP biosynthesis *de novo*—section 6.3) or that the

drugs actually block the development of the parasite at schizogony. This point can only be clarified when clean preparations of trophozoites and schizonts have been isolated and studied biochemically (section A.4).

There is evidence from studies of its antimalarial activity that menoctone (figure 9.6), the naphthoquinone currently undergoing veterinary trials for east coast fever, also blocks electron transport down the respiratory chain, probably by acting as an analogue of ubiquinone (figure 9.6 and section 4.2).

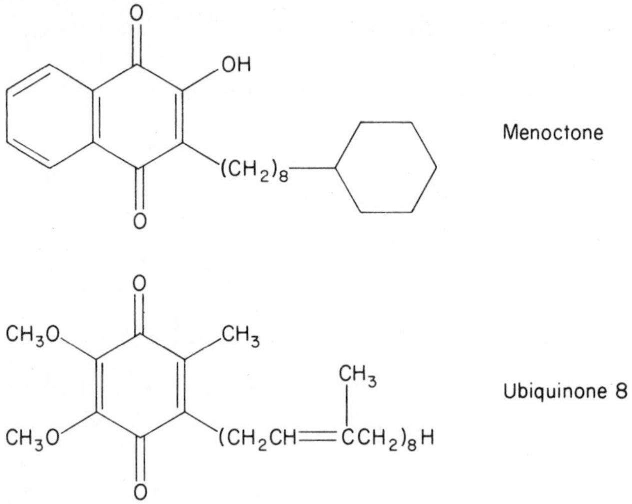

Figure 9.6 Chemical structures of menoctone and ubiquinone 8 (ubiquinone 7 has seven five-carbon isoprenoid units, ubiquinone 9 has nine and so on).

As with the anticoccidial quinolones, it is not clear whether such a blockage disrupts energy metabolism since mainly this involves only glycolysis (section 4.8) but it certainly will block other reactions which utilise the respiratory chain such as dihydroorotate dehydrogenase (section 6.3). Avian malaria parasites synthesise ubiquinones 8 and 9 and mammalian malaria parasites ubiquinones 7 and 8 (figure 9.6). This contrasts with mammalian cells, which mainly synthesise ubiquinone 10, and thus provides a possible mechanism for the selective action of the drug. No studies have yet been made of the action of menoctone on *Theileria* but it is reasonable to speculate that its mode of action and mechanism of selective toxicity are similar to those associated with its antimalarial activity.

There is evidence from electron microscopic studies that the antimalarial 8-aminoquinolines such as primaquine (figure 9.7) become concentrated within the malaria mitochondria and that they cause these organelles to swell in a manner that is reminiscent of that caused by the naphthoquinones. 8–Aminquinolines are believed to be metabolised *in vivo* to a 5,6-quinoline diquinone (figure 9.7) which has structural analogies to naphthoquinones

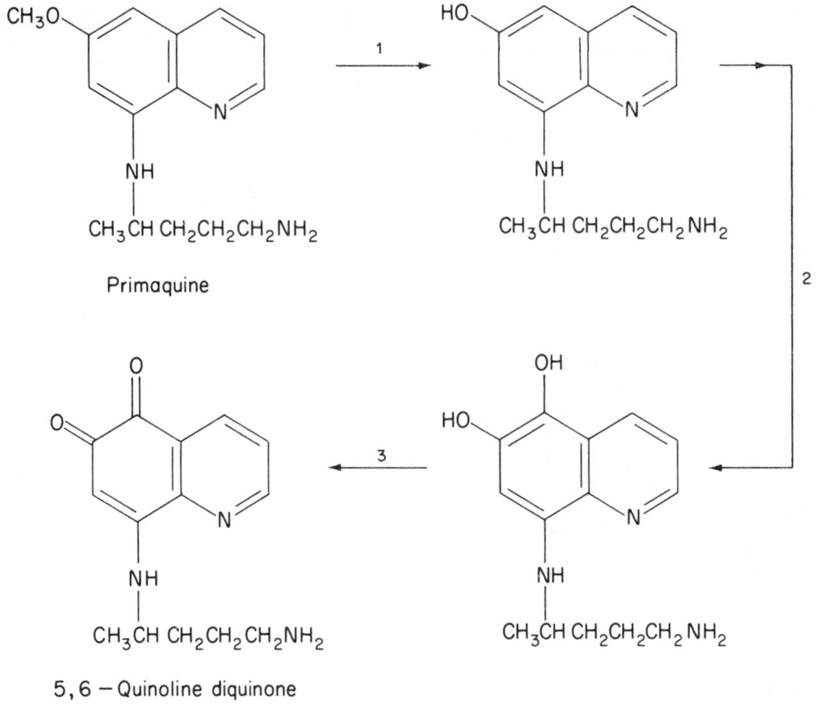

Figure 9.7 Chemical structure of primaquine and a possible route for its metabolism. Reactions: 1, demethylation; 2, oxidation; 3, rearrangement.

(figure 9.6) and the anticoccidial quinolones (figure 9.5). It has been suggested, therefore, that they too are ubiquinone analogues which act by disrupting the respiratory chain of the parasite. There is little direct evidence for this hypothesis so far, but it does seem more reasonable than the alternative hypothesis of binding to DNA, although there is evidence that this can occur *in vitro* to the purified macromolecule.

Thus, it is possible that quinolones, naphthoquinones and metabolites of 8-aminoquinolines can all disrupt the respiratory chains of parasitic protozoa, probably by acting as analogues of ubiquinones.

9.3 Drugs Interfering with Membrane Function

No antiprotozoan drug has been shown to interfere with the functioning of either the plasma membrane or lysosomal membranes of the cell but, by analogy with its antifungal action, the polyene antibiotic amphotericin B (figure 9.8), which is used in leishmaniasis, may well disrupt plasma membrane function by binding to membrane sterols and thus rendering them leaky to cations. It is possible that monensin (figure 9.8), an anticoccidial ionophoric antibiotic

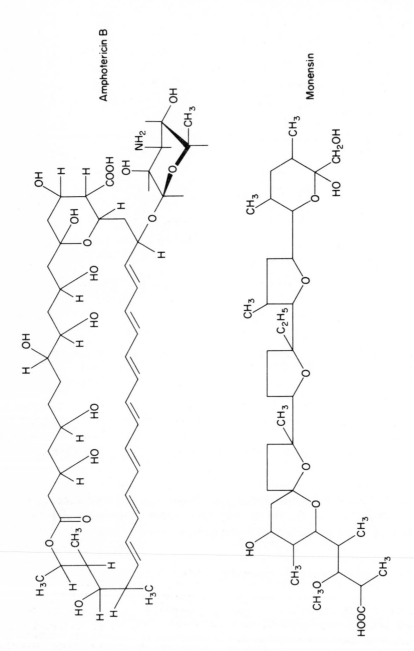

Figure 9.8 Chemical structures of amphotericin B and monensin.

acts in a similar way. However, it has also been shown that it can interact with alkali metal cations and specifically inhibit the transport of K^+ into rat liver mitochondria. It is possible, but not proven, that coccidial mitochondria may be more susceptible than those of the host to such effects and this may explain the selectivity of the anticoccidial effect of the compound.

9.4 Drugs Interfering with Cofactor Synthesis

Drugs interfering with cofactor synthesis include amprolium, the sulphonamides, sulphones, 2-substituted p-aminobenzoic acids and the 2,4-diaminopyrimidines.

The anticoccidial drug, amprolium, is a structural analogue of the dietary vitamin, thiamine (figure 9.9). Thiamine pyrophosphate is a cofactor of a

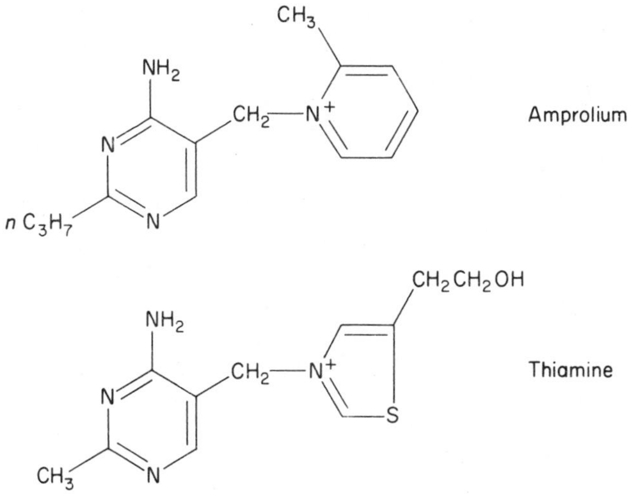

Figure 9.9 Chemical similarity of the vitamin thiamine to the anticoccidial drug amprolium.

number of decarboxylase enzymes and plays a vital role in intermediary metabolism. Because it lacks the hydroxyethyl group, amprolium cannot be pyrophosphorylated and thus it is assumed it does not inhibit at the coenzyme level. It is thought to act by inhibiting the uptake of thiamine by the parasite. The exact nature of the selectivity of the action is not clear but it is likely that the greater rate of metabolism of the parasite compared to the host contributes to this selectivity, together with a greater sensitivity of the coccidial transport system.

Sulphonamides such as sulphaquinoxaline and sulphadimethoxine, and 2-substituted p-aminobenzoic acids such as ethopabate (figure 9.10) are widely used, not only in the chemotherapy of coccidiosis, toxoplasmosis and malaria but also in the chemotherapy of bacterial diseases. Studies with bacteria have

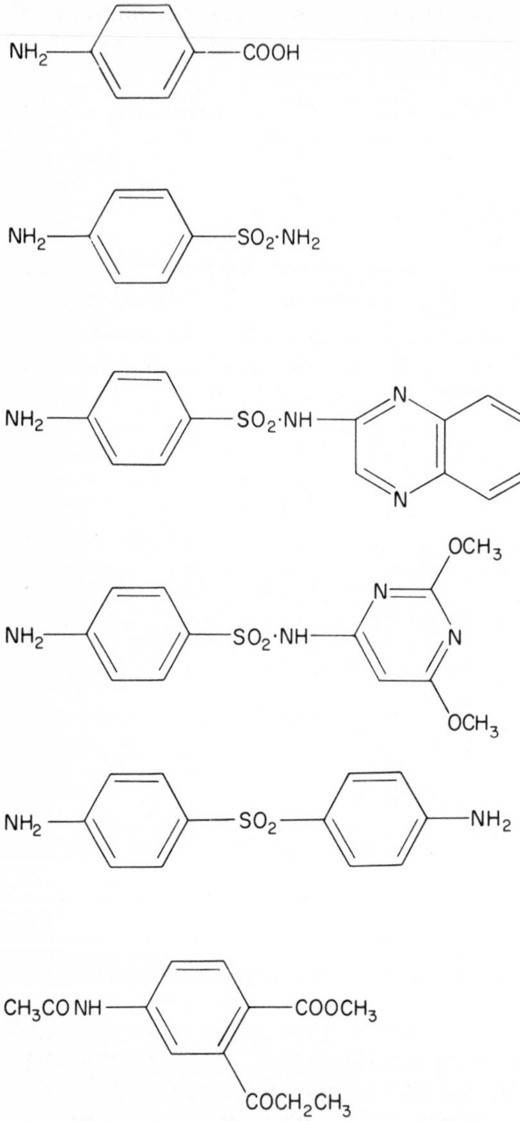

p-Aminobenzoic acid

Sulphanilamide

Sulphaquinoxaline

Sulphadimethoxine

Dapsone

Ethopabate

Figure 9.10 Chemical similarity of the growth factor *p*-aminobenzoic acid to sulphonamides, sulphones and 2-substituted *p*-aminobenzoic acids.

shown that they or their metabolites are analogues of the growth factor *p*-aminobenzoic acid (figure 9.10) and that they act by blocking the synthesis of tetrahydrofolate (section 6.4), a cofactor required, directly or indirectly, for many cellular methylation reactions (figure 9.11). Inhibition is thought to occur at the dihydropteroate synthetase reaction during which a pteridine moiety and *p*-aminobenzoic acid are conjugated (figure 9.12). It was thought originally that

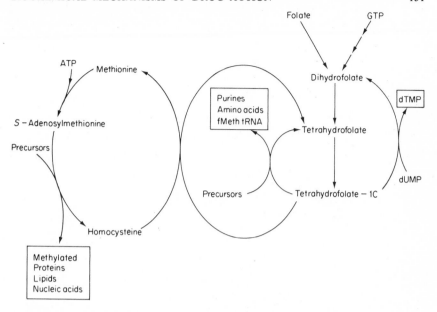

Figure 9.11 Metabolism involving the cofactor tetrahydrofolate. fMethtRNA, formyl-methionyl transfer RNA (bacteria only).

sulphonamides and sulphones were simply competitive inhibitors of this reaction but recent work has shown that in fact drug is conjugated with the pteridine to form a spurious dihydropteroate. Work on the dihydropteroate synthetase isolated from a malaria parasite has provided evidence recently that antiprotozoan sulphonamides and sulphones work in the same way as the antibacterial drugs. The selectivity of their action is easily explained since whereas the protozoa synthesise their tetrahydrofolate from GTP as illustrated in figures 9.12 and 9.13, vertebrates use preformed folate obtained in the diet (figure 9.13) and therefore do not contain a dihydropteroate synthetase.

2,4-Diaminopyrimidines such as diaveridine, ormetoprim and pyrimethamine (figure 9.14) are also used widely in the chemotherapy of coccidiosis, toxoplasmosis and malaria. Their anticoccidia and antimalaria activities have been shown to be a consequence of interference with the synthesis of tetrahydrofolate, this time at the dihydrofolate reductase reaction (see figures 9.12 and 9.13). Selectivity cannot be explained on the basis of alternative pathways of tetrahydrofolate synthesis since dihydrofolate reductase is common to both pathways. In fact, the sporozoan dihydrofolate reductases are markedly more sensitive to inhibition by some 2,4-diaminopyrimidines than are the isofunctional vertebrate enzymes (table 9.3). The biochemical basis for the differential sensitivity is not yet fully established, though it has been shown that the malaria enzyme has a higher apparent molecular weight (200 000 compared with 20 000 daltons) and different cofactor requirements to the mammalian enzyme. However, other 2,4-diaminopyrimidines **discriminate** between bacterial

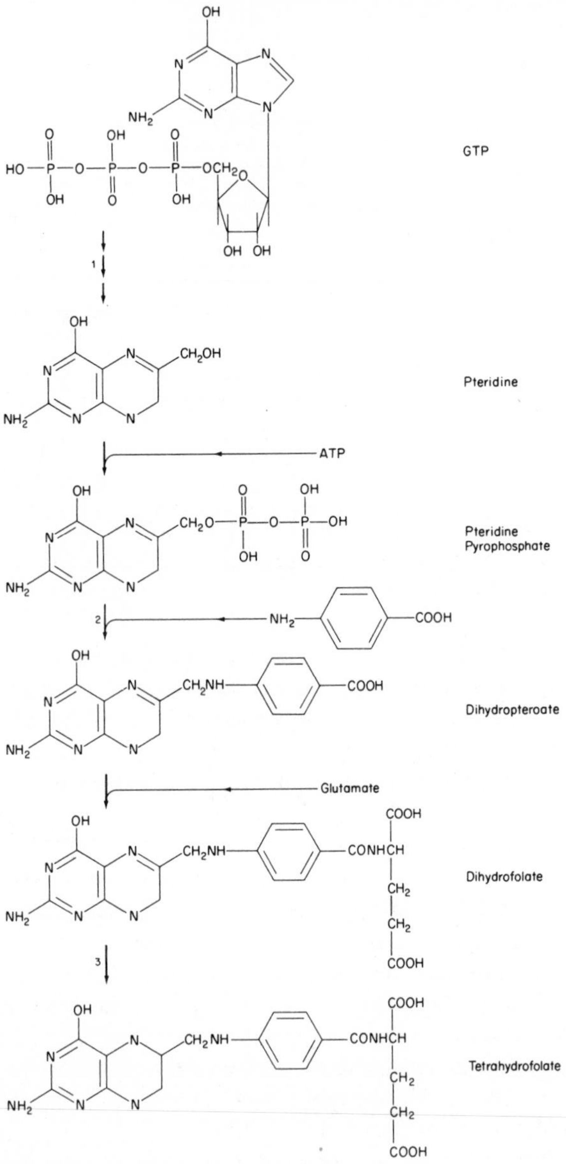

Figure 9.12 *De novo* biosynthesis of tetrahydrofolate. Enzymes: 1, GTP cyclohydrolase; 2, dihydropteroate synthetase; 3, dihydrofolate reductase.

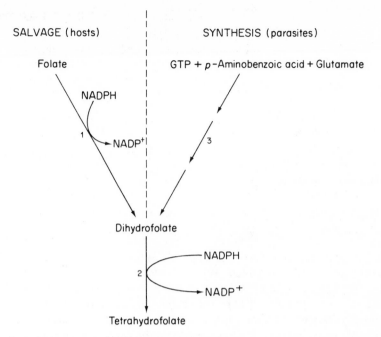

Figure 9.13 Alternative sources of tetrahydrofolate in protozoan parasites and their vertebrate hosts. Enzymes: 1, folate reductase; 2, dihydrofolate reductase; 3, series of enzymes—see figure 9.12.

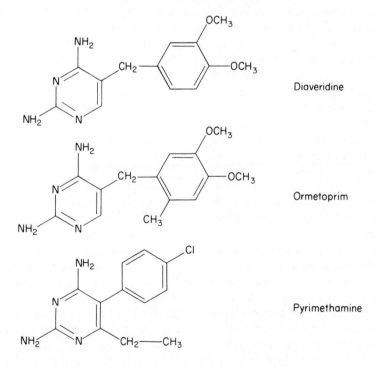

Figure 9.14 Chemical structures of antiprotozoan 2,4-diaminopyrimidines.

Table 9.3 Sensitivity of sporozoan dihydrofolate
reductases to inhibition by pyrimethamine

Parasite	Pyrimethamine concentration $(\times 10^8$ M) for 50% inhibition
Plasmodium knowlesi	0.1
Human erythrocyte	180
Plasmodium berghei	0.05
Mouse erythrocyte	100
Plasmodium lophurae	0.6
Duck erythrocyte	19
Eimeria tenella	0.3
Chicken liver	3.6

and mammalian dihydrofolate reductases which do not have such differences.

Synergism occurs between sulphonamides or sulphones and 2,4-diaminopyrimidines. One explanation of the biochemical basis for this relates to the fact that both block the same biosynthetic pathway. 2,4-Diaminopyrimidines are competitive inhibitors of dihydrofolate reductase. With inhibitors of this type, the lower the substrate concentration, the lower the concentration of drug required to produce a given degree of inhibition. Thus, it is argued that sulphonamides, by blocking the synthesis of dihydropteroate, lower the cellular level of dihydrofolate and thus potentiate the 2,4-diaminopyrimidines. An alternative explanation for this synergism involves an additional inhibitory effect of sulphonamides directly on the dihydrofolate reductase. This has been demonstrated *in vitro* but it is not clear whether the drug concentration required is achieved *in vivo*.

The biochemical consequence of interfering with tetrahydrofolate synthesis in bacteria, whether at the dihydropteroate synthetase or dihydrofolate reductase step, is disruption of active one-carbon transfer reactions (mostly methylations, hydroxymethylations and formylations) which are important in the biosynthesis of some amino acids including N-formylmethionine, purines, thymidylate, and S-adenosylmethionine (and hence other methylations) (see figure 9.11). Action at the dihydrofolate reductase level has a more rapid effect if dTMP synthesis is occurring since the thymidylate syntase reaction will actually deplete the tetrahydrofolate pool; this is normally regenerated by dihydrofolate reductase (section 6.4). Just which active one-carbon transfer reactions actually occur in sporozoan protozoa is not known, but it is unlikely that a list would include amino acid or purine biosynthesis. The thymidylate synthase reaction occurs in all genera and thus it has been suggested that the main consequence of a blockage of dihydropteroate synthetase or dihydrofolate reductase in the sporozoa is disruption of dTMP and hence DNA synthesis. There is little experimental evidence for this conclusion at present and some against it; furthermore it is unlikely that the only active one-carbon transfer reaction in the sporozoa is the thymidylate synthase reaction.

9.5 Drugs Interfering with Nucleic Acid Synthesis

Studies on antibacterial drugs have indicated that nucleic acid synthesis can be disrupted either during purine and pyrimidine synthesis or salvage or alternatively during the actual polymerisation of nucleotides into nucleic acids. No groups of antiprotozoan drugs appear to interfere with purine and pyrimidine synthesis or salvage. Studies with antibacterial drugs have also shown that inhibitors of nucleic acid synthesis which block the polymerisation step fall into two classes: those which interact with the DNA primer and thus block both

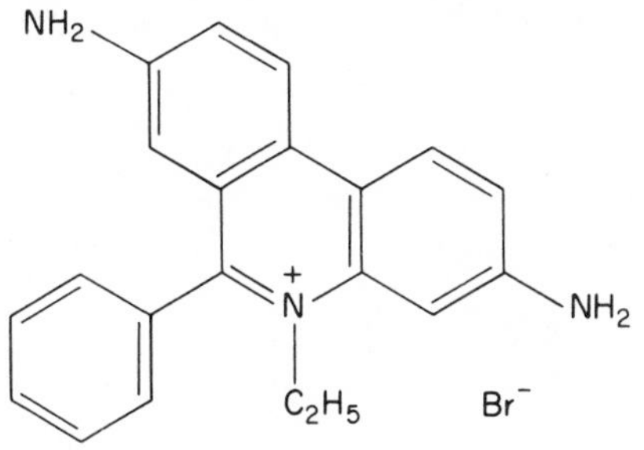

Ethidium bromide

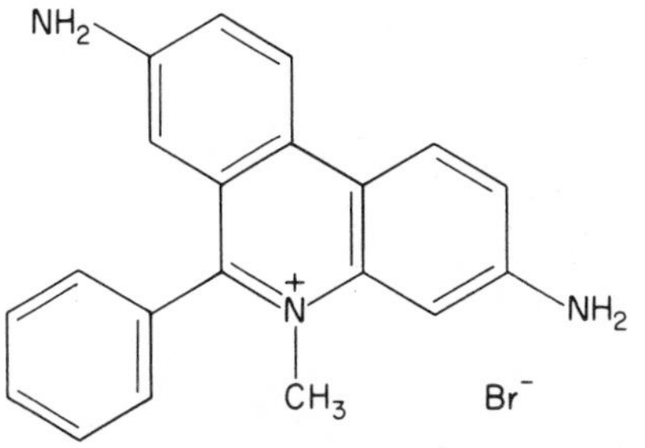

Dimidium bromide

Figure 9.15 Chemical structures of antitrypanosomal phenanthridines.

DNA and RNA synthesis; those which inhibit the polymerase proteins and thus block selectively either DNA or RNA synthesis. A large number of antiprotozoan drugs appear to block nucleic acid synthesis at the polymerisation step; all appear to do this by interacting with the DNA primer.

The antitrypanosomal phenanthridines such as dimidium and especially ethidium bromide (figure 9.15) have been most studied. Initial work with intact trypanosomes showed that ethidium inhibited selectively both DNA and RNA synthesis. Later work showed that isolated DNA polymerases and DNA-primed RNA polymerases were sensitive to inhibition and that this was the result of the intercalation of the drug with DNA. This causes a local unwinding and lengthening of the DNA helix and thus interferes with its function as a primer in nucleic acid synthesis. It seems likely that quinine, 4-amino-quinolines such as chloroquine and the quinoline and phenanthrene methanols (all antimalaria drugs—figure 9.16) all act in the same way as the phenanthridines, although it has been claimed that with chloroquinine such actions are secondary and that the primary effect concerns inhibition of haemoglobin degradation causing amino acid starvation. There is also evidence that 5-nitrofurans such as lampit, 2-nitromidazoles such as

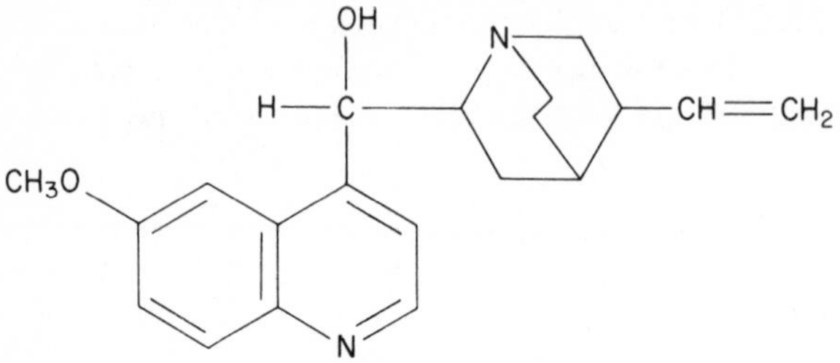

Quinine

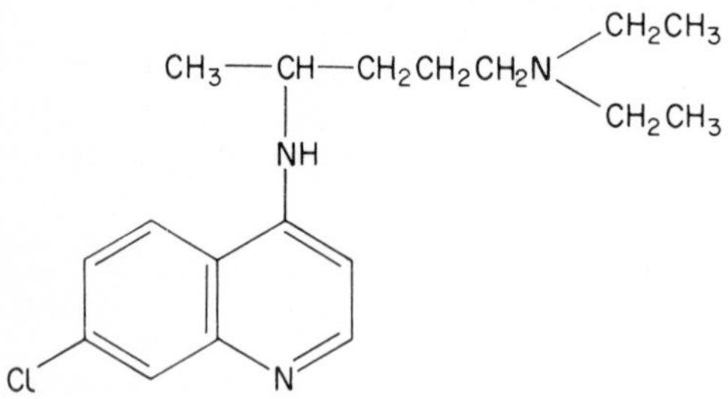

Chloroquine

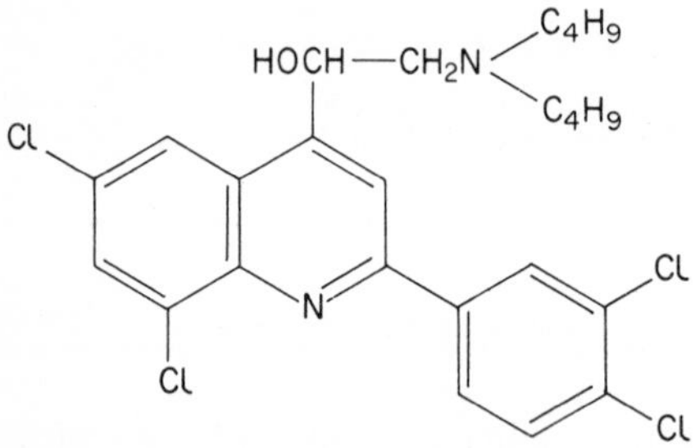

WR 30090

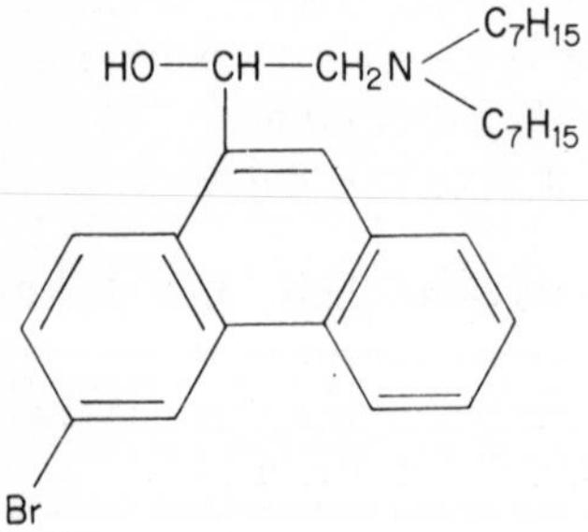

WR 33063

Figure 9.16 Chemical structures of possible DNA intercalating drugs with antiprotozoan activity.

Radanil (both active in Chagas' disease) and 5-nitroimidazoles such as metronidazole (active in trichomoniasis and amoebiasis) (figure 9.17), after suitable drug metabolism (see below) may interfere with nucleic acid synthesis by binding to DNA. It is unlikely that intercalation is involved. Similarly, there

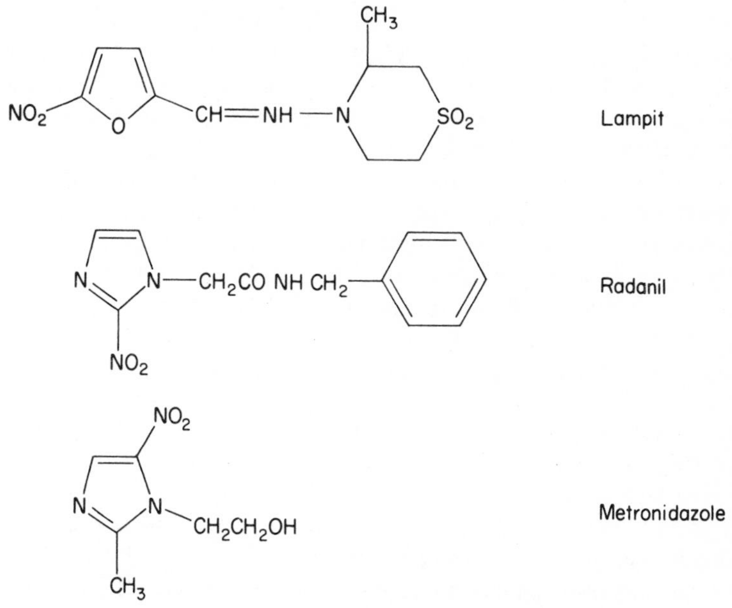

Figure 9.17 Chemical structures of antiprotozoan nitroheterocyclic drugs.

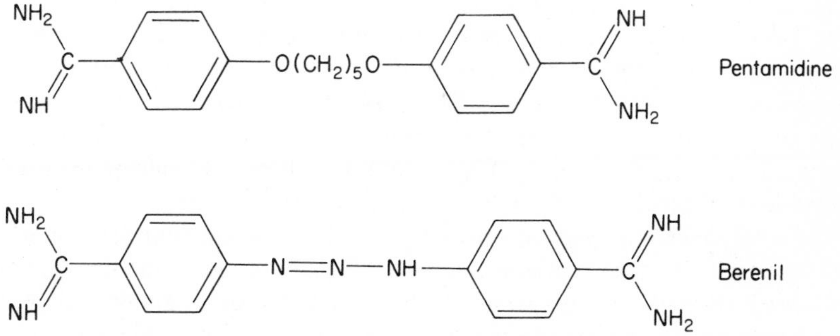

Figure 9.18 Chemical structures of antitrypanosomal aromatic diamidines.

is some evidence that the antitrypanosomal aromatic diamidines such as pentamidine and berenil (figure 9.18) can bind to DNA, although here also it is unlikely that intercalation is involved and it must be remembered that processes such as protein and phosphoglyceride synthesis are disrupted at similar concentrations to those which inhibit nucleic acid synthesis.

In addition to the general effects on nucleic acid synthesis described above, phenanthridines and diamidines, but not 5-nitrofurans, 5-nitroimidazoles and 2-nitroimidazoles, seem to exert a very specific effect on kinetoplast DNA synthesis at lower drug concentrations (see section 6.7). The reason for the specificity is not clear. It might be related to the circularity of much of the kinetoplast DNA or to its lack of association with histone; there is no firm evidence on this point. It is unlikely, however, that this interaction is related to the antitrypanosomal action of the drugs, since it is improbable that kinetoplast DNA is transcribed and translated in the blood stages. It should, however, render parasites non-infective to tsetse flies.

All the inhibitors of nucleic acid synthesis listed above can bind well to DNA from whatever source. Thus it is clear that the selectivity of their action is not a function of their ability to discriminate between DNAs of diverse sources. With most of them including ethidium bromide, dimidium, pentamidine, berenil, quinine, chloroquine, WR 30090 and WR 33063, selectivity seems to be a function of permeability: the drug penetrates into the protozoan but not into the host cells. This type of mechanism has been studied in detail only with chloroquine and pentamidine.

Chloroquine sensitive malaria parasites have at least three types of binding site for drug with apparent K_m values of about 10^{-8}, 10^{-5} and 10^{-3} M, respectively. The high affinity site, which has been shown to be membrane localised, is able to raise the intracellular concentration of drug 100–1000 times higher than that in the surrounding medium (this is usually about 10^{-6} M). The exact locations of the lower affinity binding sites are not known but they appear to occur within the cytoplasm and are probably not connected with drug uptake.

The pentamidine transport system in *Trypanosoma brucei* has been shown to be carrier-mediated and to have specificity for the aromatic amidine moiety of the drug. It has a high affinity for drug with an apparent K_m value of 2.68 μM. It is concentrative, raising the intracellular concentration of drug to many times greater than that in the external environment. The inhibitory effect of SHAM (see section 4.3) suggests that uptake is an energy-coupled process. Long slender blood trypomastigotes take up the drug much more than do short stumpy forms; culture forms hardly take it up at all, thus explaining their well-known lack of sensitivity to inhibition by the drug.

The mechanism of selective toxicity of the 5-nitroimidazoles, such as metronidazole, is more subtle and involves the activation of the drug by the partial reduction of the nitro group, probably to the hydrosylamine derivative, by the reduced electron transport protein similar to ferredoxin which is important in the terminal energy metabolism of susceptible organisms (section 4.5). This drug metabolism keeps the intracellular concentration of unchanged drug low and thus generates a concentration gradient which drives the uptake of drug by simple diffusion to levels such that the intracellular concentration of drug metabolite exceeds that of unchanged drug in the environment by a factor of 50–100. Mammalian cells do not have an enzyme complex which activates 5-nitroimidazoles and thus the mechanism of selective toxicity is explained. It is possible that a similar selective activation and uptake of 5-nitrofurans and 2-nitroimidazoles occurs, though the enzyme systems involved have not been well characterised and an explanation for selectivity is less obvious since the drugs are less electronegative than 5-nitroimidazoles and can therefore be reduced by mammalian cells.

9.6 Drugs Interfering with Protein Synthesis

The antitrypanosomal aminoquinaldine, antrycide (figure 9.19), seems to act by displacing Mg^{2+} and polyamines from the cytoplasmic ribosomes of trypanosomes. This results in their aggregation and inactivation and thus blocks protein synthesis. Such aggregation has actually been observed in trypanosomes taken from animals treated with the drug. *In vitro*, mammalian ribosomes can be aggregated equally well as those from trypanosomes so that it is likely that the mechanism of selective toxicity involves selective permeability.

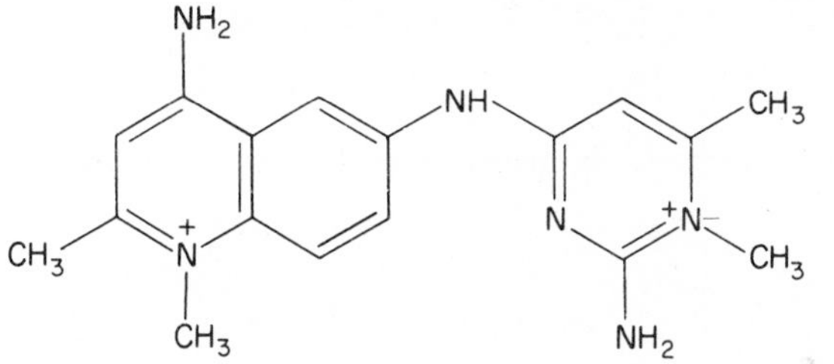

Figure 9.19 Chemical structure of antrycide.

The antibacterial action of the tetracyclines and lincomycins has been shown to be inhibition of protein synthesis. Tetracyclines block chain elongation, although the exact point of action is not yet well characterised. Lincomycins block chain elongation and the peptide bond formation step. It is likely, but not yet proven, that the antimalarial actions of the tetracycline, minocycline and the lincomycin, clindamycin (figure 9.20) are similar to those of the equivalent antibacterial drugs. The selective toxicity of the antibacterial tetracyclines results from permeability differences; that of lincomycins from differential binding to bacterial and mammalian ribosomes. Again, it is likely but not yet proven that similar mechanisms underlie the selective toxicity of minocycline and clindamycin.

9.7 Drugs with Unknown Mechanisms

There is as yet no information about the modes of action and mechanisms of selective toxicity of three of the currently used anticoccidial drugs, clopidol (a pyridone), carbanilide and zoalene (a nitrobenzamine) (figure 9.21).

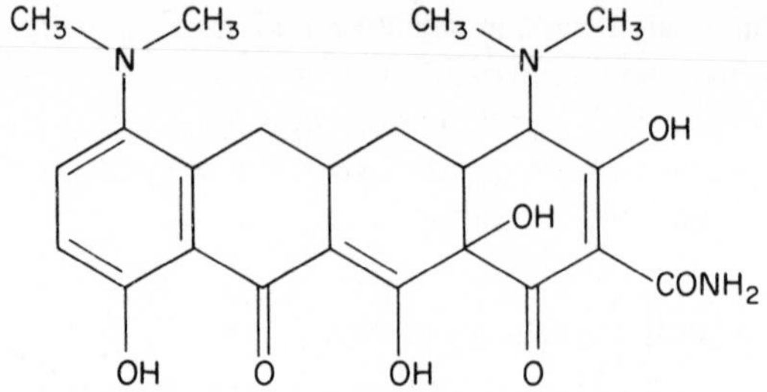

Minocycline

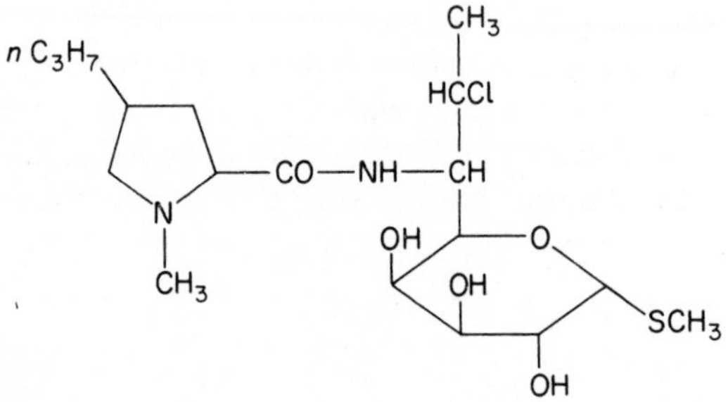

Clindamycin

Figure 9.20 Chemical structures of an antiprotozoan tetracycline and an antiprotozoan lincomycin.

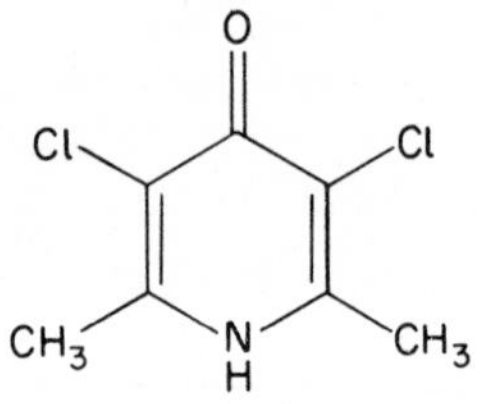

Clopidol

Carbanilide

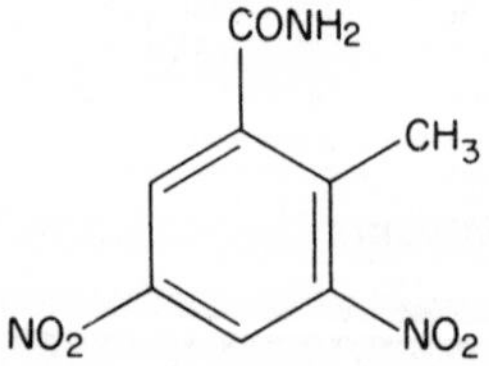

Zoalene

Figure 9.21 Antiprotozoan drugs with unknown mechanisms of action.

9.8 General Conclusions about Mechanisms of Action

The biochemical mechanisms of action of the antiprotozoan drugs are summarised in table 9.4. It is clear that the majority interfere with the biosynthetic reactions of the cell and especially with nucleic acid synthesis. Only a few disrupt membrane functioning but the list of those which might affect energy metabolism is longer than was originally suspected. It includes, however, the arsenicals and antimonials—the most toxic of the drugs on our list which are used for human infections—and a number of drugs used only for veterinary purposes.

9.9 General Conclusions about Mechanisms of Selective Toxicity

The biochemical mechanisms of selective toxicity of antiprotozoan drugs are summarised in table 9.5. Many drugs appear to be selective because of differences in permeability. There are, however, two sites of action which have no equivalent in vertebrate cells: L-α-glycerophosphate oxidase in trypanosomes and dihydropteroate synthetase in the Sporozoa. These are exploited respectively by suramin and the sulphonamides and sulphones. In addition, 5-nitroimidazoles such as metronidazole are selective because they are activated in the parasite by a system with no equivalent in mammalian cells. There are also a number of sites where drugs can discriminate between isofunctional systems in the protozoan and the vertebrate: pyruvate kinase, phosphofructokinase, thiamine uptake, K^+ uptake by mitochondria, ubiquinone, dihydrofolate reductase and at least one ribosome function. This relatively short list is undoubtedly a reflection both of the similarity of the biochemistry of protozoa and vertebrate cells and also of present poor knowledge of the mechanisms of action of many antiprotozoan drugs.

9.10 Mechanisms of Drug Resistance

Studies, particularly with antibacterial drugs, have led to the recognition of five basic mechanisms by which microorganisms can become resistant to drugs:

(1) metabolism of drug to an inactive form;
(2) alteration in permeability so drug no longer enters pathogen;
(3) metabolic lesion bypassed by alternative metabolic pathway;
(4) target altered so that its sensitivity to inhibition is decreased;
(5) target enzyme increased in quantity so that pathogen can survive high percentage inhibitions of activity.

Little experimental work has been done on mechanisms of resistance of antiprotozoan drugs though examples are known of resistance in categories (2), (4) and (5).

Table 9.4 Mechanisms of action of antiprotozoan drugs

Energy metabolism	Membrane function	Cofactor synthesis	Nucleic acid synthesis	Protein synthesis	Unknown
Tryparsamide	*Monensin	Amprolium	*Pentamidine	Antrycide	Clopidol
Mel B	*Amphotericin B	Ethopabate	Ethidium	Minocycline	Carbanilide
*Suramin		Diaveridine	·Dimidium	Clindamycin	Zoalene
Pentostam		Sulphaquinoxaline	*Berenil		
*Robenidine		Ormetoprim	Lampit		
Buquinolate		Sulphadimethoxine	*Radanil		
Decoquinate		Dapsone	*Metronidazole		
Methylbenzoquate		Pyrimethamine	Quinine		
*Menoctone			Chloroquine		
*Primaquine			WR 30090		
			WR 33063		

* Evidence not conclusive

Table 9.5 Mechanisms of selective toxicity of antiprotozoan drugs

Differential permeability	Drug activation in protozoan only	Unique target in protozoan	Drug discrimination between isofunctional targets	Pathway in protozoan more important	Unknown
Tryparsamide	*Lampit	*Suramin	Tryparsamide	Tryparsamide	Clopidol
Mel B	*Radanil	Ethopabate	Mel B	Mel B	Robenidine
Pentamidine	Metronidazole	Sulphaquinoxaline	Pentostam	Pentostam	Carbanilide
Ethidium		Sulphadimethoxine	*Amprolium		Zoalene
Dimidium		Dapsone	Buquinolate		Amphotericin B
Antrycide			Decoquinate		Monensin
Berenil			Diaveridine		
Pentostam			Ormetoprim		
Quinine			Methylbenzoquate		
Chloroquine			Pyrimethamine		
Minocycline			*Primaquine		
*WR 30090			Clindamycin		
*WR 33063			Menoctone		

* Evidence not conclusive

A very common mechanism of resistance, seen particularly in drugs with a mechanism of selective toxicity involving differential permeability, is reduction in permeability to the drug. There is at least some evidence that resistance to diamidines, phenanthridines, aminoquinaldenes and 4-aminoquinolines falls into this category. Only resistance to the 4-aminoquinoline chloroquine and the diamidine, pentamidine have been studied in detail. Resistance to chloroquine involves loss of the high affinity chloroquine-binding site and thus failure to concentrate drug from the medium. It appears to be a stable character which is inherited in a simple Mendelian fashion, undergoes genetic recombination with other markers and probably arises by mutation and selection in the presence of drug. Resistance to pentamidine appears to involve alterations in the transport system such that it has either lower V_{max} or higher K_m or a combination of the two.

Strains of malaria resistant to pyrimethamine have been shown to contain increased quantities of an altered dihydrofolate reductase which no longer binds the drug as well as the parent enzyme (table 9.6). Resistance arises by mutation and the genetic factors involved can undergo recombination with other markers in crosses between resistant and sensitive parasite lines.

Table 9.6 Properties of dihydrofolate reductase from strains of malaria parasite (*Plasmodium berghei*) sensitive and resistant to pyrimethamine

Strain	Specific activity*	K_m for dihydrofolate†	ID_{50}‡
Sensitive	1.6	3.5	1.0
Resistant	5.0	30	30

* nmol/min/mg protein
† $\times 10^{-6}$ M
‡ Concentration of drug ($\times 10^8$ M) for 50 per cent inhibition

It is likely that further work in this neglected area of protozoan biochemistry will yield examples of other mechanisms. Such work is to be encouraged as investigation of the mechanism of resistance of a drug often gives additional information on its mechanism of action.

9.11 General Conclusions

The principal reason for studying the biochemical modes of action and mechanisms of selective toxicity of drugs is the hope that information derived from the studies will make it possible for a new generation of antiprotozoan drugs to be developed much more rationally than has been possible up to now. Clearly, for this to happen a more detailed understanding of drug action than is available at present will be required, but nevertheless the way forward is clear and many more studies are known to be in progress. Included are investigations of drug permeability. It is clear from table 9.5 that a common mechanism of selective toxicity involves differential drug uptake. This can be effective even with intracellular parasites, provided the host cell is permeable to the drug and the parasite can concentrate it (for example, chloroquine and malaria). Thus a more detailed understanding of the mechanisms behind differential drug permeability could be invaluable for drug development in the future.

9.12 Further Reading

Coombs, G. H. (1976). Studies on the activity of nitroimidazoles. In *Biochemistry of Parasites and Host–Parasite Relationships* (ed. H. Van den Bossche), Elsevier-North Holland Biomedical Press, Amsterdam, pp. 545–552

Franklin, T. J. and Snow, G. A. (1975). *Biochemistry of Antimicrobial Action*. Second edition, Science Paperbacks, Chapman and Hall, London

Gale, E. F., Cundliffe, E., Reynolds, P. E., Richmond, M. H. and Waring, M. J. (1972). *The Molecular Basis of Antibiotic Action*. John Wiley and Sons, London

Newton, B. A. (1974). The chemotherapy of trypanosomiasis and leishmaniasis: towards a more rational approach. In *Trypanosomiasis and Leishmaniasis with Special Reference to Chagas' Disease*. Ciba Foundation Symposium 20 (new series). Associated Scientific Publishers, Amsterdam, pp. 285–307

Pinder, R. M. (1971). Recent advances in the chemotherapy of malaria. *Progr. med. Chem.*, **8**, 231–316

Ryley, J. F. and Betts, M. J. (1973). Chemotherapy of chicken coccidiosis. *Adv. Pharmac. Chemother.*, **11**, 221–293

10 Present Position and Future Prospects

A major reason for investigating the biochemistry of parasitic protozoa is the potential that the results should provide for the rational approach to the chemotherapy of many important protozoan diseases. Most biochemical studies of parasitic protozoa are directed towards this final goal. However, it is clear that during the process of such investigations many new and exciting discoveries have been made, which, although not immediately applicable, may finally prove of immense value.

A rational approach to chemotherapy is dependent on a good knowledge of the biochemistry of both parasite and host. Mammalian biochemistry still contains many mysteries, but we know far less about parasitic protozoa. The reasons for this are clear. Protozoa are not easy organisms with which to work; to obtain large quantities of uncontaminated material is impossible for many species, difficult for most. Many species will not grow outside of their host. This does not affect studies of the host–parasite relationships, but makes investigations of the protozoa very difficult. It is easy to understand why most biochemists work with bacteria and rat liver. A further problem is finance. Most protozoan diseases are not endemic in the developed, wealthy countries of the world but rather inflict the poor of the developing countries. It is not economically viable therefore for pharmaceutical companies to invest large sums of money in drug development programmes in this area and, consequently, most of them have now stopped screening for compounds active against these diseases.

Despite these problems, progress is being made. Our understanding of the biochemistry of parasitic protozoa is increasing, even if slowly. Already many exciting and potentially exploitable systems have been discovered. Included here are the hydrogenosomes of *Trichomonas*, the L-α-glycerophosphate oxidase system of trypanosomes, kinetoplast DNA of the Kinetoplastida, pyrophosphate-linked pathways in *Entamoeba*, variant antigens in trypanosomes and polyhistidine in *Plasmodium*. These are just a few of the areas which have provided some exciting findings in recent years. Clearly this is just the beginning, especially as the progress being made with culture and separation techniques should ease the major problem of availability of material.

Where do we go from here? A three-part approach is required. Many more metabolic studies are needed to unravel the biochemical pathways that occur in parasitic protozoa and to identify further potential targets for chemotherapeutic attack. In addition, more research needs to be carried out on the biochemical modes of action of existing drugs and on mechanisms of resistance since such studies not only help to elucidate the metabolism of the organism but also provide information which will be vital to future attempts at more rational drug therapy directed towards specific target sites. As well as these biochemical studies it is essential that the conventional method of finding new drugs, the screening of compounds against parasites *in vitro*, or in animals, should continue. Information from the other two approaches, as it becomes available, should enable an increasingly more rational approach to be taken in the selection of compounds to be synthesised and tested.

A major difficulty clearly is going to be finance. Money for most of this work will have to come from national and international agencies. It is to be hoped that the World Health Organization's plans for an ambitious programme of research and training aimed at eradicating six major tropical diseases (including trypanosomiasis, leishmaniasis and malaria) will come to fruition and provide a welcome injection of funds into this neglected area of research.

The problems are clear, the rewards obvious. This is an exciting field indeed for a biochemist, a field in which major discoveries are still to be made, and one with an immensely important application. We feel that one factor responsible for the relatively few biochemists working on parasitic protozoa is that many are completely uninformed as to the interests and problems involved. We hope that this book will help to change this situation.

Appendix A Isolation of Parasitic Protozoa from Infected Animals

A.1 Genus *Trypanosoma* (except *T. cruzi*)

Before 1968, blood trypomastigotes for biochemical studies were usually isolated by simple differential centrifugation. This procedure often can yield preparations uncontaminated by host red and white blood cells but large numbers of platelets (up to 30 per cent by number) will be present. These cell fragments contain many enzyme activities such as the entire glycolytic sequence including lactate dehydrogenase (see chapter 4). When the platelets are removed using the technique of defibrination (which essentially involves allowing the blood to clot), the lactate dehydrogenase activities reported previously to be present in trypanosomes such as *T. rhodesiense* can no longer be detected. The consequence of this discovery is that the literature prior to 1968 on enzymatic studies of trypanosomes is treated with suspicion, unless the infected blood was defibrinated before centrifugation. Protozoan biochemists now take much more notice of the importance of working with pure preparations of parasites.

The current method of choice involves the use of columns of DEAE cellulose swollen in either a phosphate–saline–glucose (PSG) or Tris–saline–glucose (TSG) buffer of suitable pH and ionic strength (for example, PSG of pH 8.0, $I = 0.217$ for *T. brucei*). The swollen cellulose is poured into a scintered glass funnel and the column of cellulose thus formed is thoroughly washed in PSG or TSG. Infected blood, collected with heparin as anticoagulant, is poured on to the top of the column. When the blood has completely penetrated into it, the column is eluted with PSG or TSG. Red and white blood cells and platelets are adsorbed on to the cellulose; trypanosomes come through in the eluate essentially uncontaminated by host blood cells and can be collected by centrifugation.

The factors which most influence the degree of separation achieved are the buffer (PSG or TSG) and its pH and ionic strength. The surface charges of both trypanosomes and mammalian host blood cells vary and so the optimum conditions for separation have had to be determined for each host–parasite combination. Recoveries of up to 100 per cent have been recorded with some species of trypanosomes and values rarely fall below 50 per cent. Highest recoveries are obtained when there is the greatest difference in surface charge between host blood cells (especially red cells) and parasites. Species isolated by this method are (in order of ease of separation): *T. simiae* (a parasite of pigs), *T. evansi, T. gambiense, T. rhodesiense, T. brucei, T. congolense, T. vivax* and *T. lewisi.* Trypanosomes isolated from DEAE cellulose columns are motile, normal in appearance and have lost none of their infectivity for their hosts.

None of the stages parasitic in the insect vector has yet been isolated in quantities sufficient for biochemical study although some preliminary work has been carried out with small numbers of mid-gut forms.

A.2 *Trypanosoma cruzi*

Blood trypomastigotes can be isolated by a DEAE cellulose column method similar to that used for other trypanosomes provided most of the red cells are first removed by differential centrifugation ($500g$ for 5 min). The pH of the eluting buffer is critical (PSG, pH 7.5, $I = 0.206$) and organisms need to be collected in serum since they are very friable. Recoveries are about 50 per cent.

Preparation of the intracellular amastigote stages is more difficult. These have now been isolated from the hind skeletal muscle of rats, mice and chinchillas. The tissue is disrupted by homogenisation to release the parasites. The homogenate is then filtered through gauze to remove large debris and incubated with DNase, to destroy muscle cell nuclei, collagenase, to break down collagen-like proteins, and trypsin, to degrade other proteins. Parasites are then collected from the homogenate by differential centrifugation and further purified, if necessary, in a low-speed ($225g$ for 5 min) linear sucrose gradient (0.25–0.70 M sucrose). Recoveries range from 30 to 70 per cent, contamination from 0.2 to 2.0 per cent, depending on whether a sucrose gradient step is used. Stages parasitic in the reduviid bugs have not yet been isolated.

A.3 Genus *Leishmania*

Leishmania donovani amastigotes have been isolated from spleen macrophages, though detailed methodology has yet to be published. The spleen is removed, homogenised in a phosphate-buffered saline solution and the amastigotes purified by differential centrifugation. *L. mexicana* amastigotes have been isolated from mouse lesions by coarse homogenisation, passage through tissue paper and washing in tissue culture medium 199. Promastigotes have not been isolated from sand flies.

A.4 Genus *Eimeria*

No-one has yet isolated trophozoites and schizonts from chicken intestinal epithelial cells. Methods for preparation from tissue cells have been published but have yet to be used in biochemical studies. In contrast, oocysts can be easily extracted and purified from the faeces of infected chickens. This is why most of the biochemical studies of *Eimeria* have been carried out with these stages. Merozoites and sporozoites also have been isolated.

A.5 Genus *Toxoplasma*

Trophozoites have been isolated from peritoneal macrophages. The macrophages are collected by centrifugation from peritoneal washings and shaken with glass beads to release the parasites. White cells are removed by passage of the homogenate through a sintered glass funnel or by density gradient centrifugation and red cells by immunolysis. After a thorough washing parasites are ready for use. None of the other stages has yet been isolated in quantities sufficient for biochemical study.

A.6 Genus *Plasmodium*

Intraerythrocytic stages can be isolated readily from infected blood. Since infection in many species develops synchronously, individual stages within the intraerythrocytic cycle can be purified by collecting the blood at the appropriate

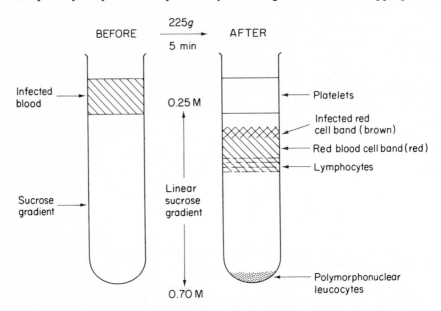

Figure A.1 Isolation of malaria-infected red blood cells using a linear sucrose gradient.

time. The methodology used involves first removing white cells, platelets and as many uninfected red cells as possible and then releasing the parasites by disruption of the infected red cells. White cells are often removed by passage of the blood through columns of dry cellulose powder. Another method is to use a low speed (225g for 5 min) linear sucrose gradient (0.25–0.70 M) which separates the infected red cells not only from the white cells and platelets but also from most of the non-infected red cells (figure A.1). The infected red cell band can be distinguished easily by its brown colour (due to malaria pigment—see section 1.10) and can readily be removed. Parasites can be released from red cells by a number of methods including saponin lysis, nitrogen cavitation and disruption in the French press. The most gentle method involves immunolysis with anti-red cell serum in the presence of an excess of complement. It must be remembered, however, that none of these methods removes adequately the erythrocyte-derived plasma membrane by which the parasite becomes surrounded during penetration into the red cell. After immunolysis, parasites are collected and washed thoroughly by centrifugation. Recoveries are greater than 50 per cent and contamination by white cells is less than 0.1 per cent.

Merozoites have been obtained by incubating infected blood in a culture chamber fitted with a polycarbonate sieve of 2 μm pore size. This allows merozoites (but not infected blood cells) to pass out of the chamber as they are released. Exoerythrocytic liver schizonts have not yet been isolated, but oocysts from mosquito gut and sporozoites from mosquito salivary glands have been obtained in small quantities.

A.7 Further Reading

Lanham, S. H. and Godfrey, D. G. (1970). Isolation of salivarian trypanosomes from man and other mammals using DEAE cellulose. *Expl Parasit.*, **28**, 521–534

Williamson, J. and Cover, B. (1975). The rapid isolation from human blood of concentrated, white cell free preparations of *Plasmodium falciparum*. *Trans. R. Soc. trop. Med. Hyg.*, **69**, 78–87

Appendix B Culture of Parasitic Protozoa

B.1 Genus *Trypanosoma* (except *T. cruzi*)

In the past, culture forms have been grown in both diphasic (for example, blood agar base with a buffered salts–glucose overlay in which the organisms grow) and monophasic (for example, Pittam's medium: tryptose–caesin hydrolysate–liver digest–buffered salts–glucose–blood extract) media. A defined medium is, however, now available and being used extensively (Cross and Manning's medium—table B.1). It contains haemin since a characteristic feature of trypanosomatids is an inability to synthesise the porphyrin ring. Note that cultures are incubated at vector temperatures (25 °C) not mammalian temperatures (37 °C).

Blood trypomastigote stages have been cultured for up to five generations in tissue culture media in the presence of a monolayer of mammalian tissue cells. Recently, the use of tissue culture medium RPMI 1640 over a layer of fibroblasts has been reported to support serial culture of blood-stream *T. brucei* and *T. congolense* at 37°C, although only at relatively low cell densities.

B.2 *Trypanosoma cruzi*

Culture epimastigote forms can be grown at 25 °C in a number of media including LIT (tryptose–liver infusion–buffered saline–glucose–haemin–calf serum). It has been reported also that they will grow in Cross and Manning's defined medium (table B.1). Blood trypomastigote stages do not grow and divide *in vivo* without prior transformation to intracellular amastigote stages so that, not surprisingly, these cannot be cultured serially. They have, however, been produced in large numbers in tissue culture. Amastigote stages have been cultured serially in Pan's medium (tissue culture medium 199–trypticase–haemin–bovine foetal serum–chicken plasma) at 35 °C but it is not clear how similar they are to intracellular amastigote stages. The latter can be grown in tissue culture at 37 °C.

B.3 Genus *Crithidia* (except *C. oncopelti*)

These model trypanosomatids will grow readily at 25 °C in undefined (for example, peptone–liver infusion–NaCl–glucose–haemin–adenine–folic acid)

Table B.1 Composition of defined media for parasitic protozoa

	Trypanosoma (Cross and Manning, 1973)	Crithidia fasciculata (Kidder and Dutta, 1958)	Crithidia oncopelti (Newton, 1956)	Leishmania tarentolae (Trager, 1957)	Trichomonas (Shorb and Lund, 1959)	Plasmodium* (Trager, 1976)
Salts	NaCl, KH$_2$PO$_4$	NaCl, K$_2$HPO$_4$, NaH$_2$PO$_4$	NaCl, KCl, Na$_2$HPO$_4$, NH$_4$Cl	NaCl, Na$_2$HPO$_4$, KH$_2$PO$_4$	K$_3$PO$_4$	NaCl, KCl, Na$_2$HPO$_4$, MgSO$_4$, Ca(NO$_3$)$_2$
Additional buffers	HEPES NaHCO$_3$	0	0	0	0	NaHCO$_3$
Trace ions	–	–	–	10	8	–
Energy sources	Glucose, citrate, acetate, succinate, glucosamine	Glucose	Glucose	Glucose	Mannitol, inositol, glycerophosphate	Glucose
Amino acids	21	11	1	17	16	21
Purines	2	1	1	3	2	0
Pyrimidines	2	0	0	2	4	0
Fatty acids	Linoleic acid–albumin complex Tween 40	0	0	0	2	0
Sterols	0	0	0	0	1	0
Growth factors	17	7	5	12	15	11
Other components	Haemin EDTA	Haemin EDTA	0	Haemin	EDTA ascorbic acid	Glutathione 15% human serum
pH	7.4	8.0	7.4	8.0	7.0	7.4
Incubation temperature (°C)	25	25	25	25	37	37

* Not strictly a defined medium as it contains 15 per cent serum; –, not added as such; 0, none

and defined (for example, Kidder and Dutta's medium—table B.1) media. The high levels of folic acid in these media can be replaced partially by addition of an unconjugated pteridine such as biopterin. A nutritional requirement for an unconjugated pteridine is unusual among microorganisms and has led to the use of *C. fasciculata* as a biological assay for these compounds.

B.4 *Crithidia oncopelti*

This trypanosomatid will grow at 25 °C in very simple undefined (peptone–NaCl–glucose) and defined (table B.1) media without haemin. The use of *C. oncopelti* in biochemical studies of trypanosomes, however, is best avoided (see appendix C).

B.5 Genus *Leishmania*

Promastigote stages can be cultured at 25 °C in biphasic (for example blood agar base with an overlay of glucose-containing buffered salts) or monophasic (for example brain heart infusion with 10 per cent citrated, lysed, filtered human blood) media. Defined media (for example, Trager's medium—table B.1) are also available. Amastigote stages can readily be grown at 37 °C in tissue culture.

B.6 Genus *Trichomonas*

Most parasitic protozoa are aerobic. Two exceptions are *Trichomonas* and *Entamoeba* which prefer anaerobic conditions. These are obtained by the addition of reducing compounds such as ascorbic acid to the medium. The metabolism of these organisms will also tend to produce an anaerobic environment. Both undefined (for example, Diamond's medium: trypticase–yeast extract–maltose–cysteine–ascorbic acid–agar–heat inactivated sheep serum) and defined (for example, Shorb and Lund's medium, table B.1) media can be used for *Trichomonas*. Cultures are incubated at 37 °C.

B.7 Genus *Entamoeba*

Only undefined media (for example, Carter, Levy and Diamond's medium: peptone–liver digest–glucose–vitamins–serum) are available for axenic culture.

B.8 Genera *Eimeria*, *Toxoplasma* and *Theileria*

Trophozoites of *Eimeria* and *Toxoplasma* can be grown in tissue culture but these have been little used so far in biochemical studies. The exoerythrocytic stages of *Theileria* can be grown in transformed lymphoblasts and are used in experimental vaccine production. Recently, methods have been developed which support growth of all the mammalian stages.

Table B.2 Growth factors for parasitic protozoa

Growth factor	Required by	Probable function
Thiamine (vitamin B_1)	Most parasitic protozoa (not *Trichomonas*)	Precursor of thiamine pyrophosphate, a coenzyme in many decarboxylation reactions
Riboflavin (vitamin B_2)	*Crithidia fasciculata* *Leishmania tarentolae*	Precursor of FAD and FMN coenzymes in electron transfer reactions
Pyridoxine (vitamin B_6) or related compounds	*Crithidia fasciculata* *Leishmania tarentolae* *Trichomonas*	Precursor of pyridoxal phosphate, a coenzyme in many enzyme reactions, especially those involving transamination
Pantothenate	*Crithidia fasciculata* *Leishmania tarentolae* *Trichomonas*	Precursor of CoA, important in many reactions as a carrier of acetyl, malonyl and succinyl groups
CoA	*Plasmodium lophurae*	As above
Biopterin	*Crithidia fasciculata*	In hydroxylase reactions including the desaturation of fatty acids and the synthesis of orotate
Folic acid	*Crithidia fasciculata* *Leishmania tarentolae* *Trichomonas*	Precursor of tetrahydrofolate, important in active one-carbon transfer reactions
p-Aminobenzoic acid	*Plasmodium*	In the synthesis of tetrahydrofolate
Biotin	*Crithidia fasciculata* *Leishmania tarentolae*	Coenzyme in carboxylation reactions
Nicotinic acid	*Crithidia fasciculata* *Leishmania tarentolae* *Trichomonas gallinae*	Precursor of NAD(P)
Haemin	Trypanosomatias	Prosthetic group of cytochromes, catalase and peroxidase

B.9 Genus *Plasmodium*

After many years of effort, serial culture of the intraerythrocytic stages was first obtained in 1976. Best results to date have been with *P. falciparum*, one of the human malaria parasites. One medium used is a tissue culture medium designated RPMI 1640 (developed originally for human leukocytes) supplemented with HEPES buffer and 15 per cent human serum. A gas phase with 7 per cent CO_2, and a low oxygen tension (around 5 per cent) is provided. Human red blood cells infected at a low level with parasites (approximately 0.1 per cent) are held in a stationary layer and the medium is changed either daily by hand, or continuously by a slow flow method. The blood cells are diluted with fresh uninfected cells every third or fourth day. In the initial period of 50 days, the original infected cell suspension was diluted more than 10^7 times. Parasites lost their synchrony and gametocytes were not seen but some infectivity to *Aotus* monkeys was retained. It is not clear which of the particular features of this system are critical in obtaining success; most likely this is a result of selection of the correct permutation of all the possible combinations. No other stages can be cultured serially.

B.10 General Conclusions

It is clear from the previous sections that although a number of undefined media are available for some genera of parasitic protozoa, few defined media have so far been produced. The composition of those that have are summarised in table B.1. With the exception of those for *Crithidia* species they are not minimal media so that their constituents do not allow conclusions to be drawn about the overall biosynthetic capabilities of the parasites. However, all seem to require at least some growth factors. A list of the more common ones and their likely biochemical role is given in table B.2.

B.11 Further Reading

Cross, G. A. M. and Manning, J. C. (1973). Cultivation of *Trypanosoma brucei* sspp. in semidefined and defined media. *Parasitology*, **67**, 315–331

Kidder, G. W. and Dutta, B. N. (1958). The growth and nutrition of *Crithidia fasciculata*. *J. gen. Microbiol.*, **18**, 621–638

Newton, B. A. (1956). A synthetic growth medium for the trypanosomatid flagellate *Strigononas* (*Herpetomonas*) *oncopelti*. *Nature*, **177**, 279–280

Shorb, M. S. and Lund, P. G. (1959). Requirements of trichomonads for unidentified growth factors, saturated and unsaturated fatty acids. *J. Protozool.*, **6**, 122–130

Taylor, A. E. R. and Baker, J. R. (1968). *The cultivation of parasites* in vitro. Blackwell Scientific Publications, Oxford

Trager, W. (1957). Nutrition of a hemoflagellate (*Leishmania tarentolae*) having an interchangeable requirement for choline or pyridoxal. *J. Protozool.*, **4**, 269–276

Trager, W. (1976). Prolonged cultivation of malaria parasites *Plasmodium coatneyi*
and *P. falciparum*. In *Biochemistry of Parasites and Host–Parasite Relationships*
(ed. H. Van den Bossche), Elsevier-North Holland Biomedical Press, Amsterdam,
pp. 427–434

Appendix C The Enigma of
Crithidia oncopelti

C.1 The Presence of Bipolar Bodies

Investigations of the biochemistry of trypanosomes are, as was indicated in section 3.2, still seriously hampered by problems of culturing these organisms *in vitro*. One way round these difficulties is to use some of the readily cultured insect trypanosomatids (especially members of the genus *Crithidia*) as model systems. By far the easiest of these to handle is *C. oncopelti*, which has thus been used in a number of investigations.

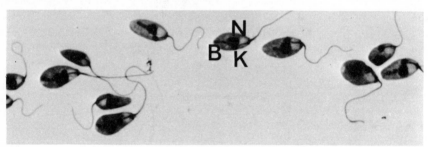

Figure C.1 Light photomicrograph of a fixed and stained preparation of *Crithidia oncopelti* showing the presence of bipolar bodies (1000). N, nucleus; K, kinetoplast; B, bipolar bodies.

Reference has already been made in appendix B to the fact that *C. oncopelti* will grow in a simple, completely defined medium, consisting of salts, glucose, methionine, adenine and five growth factors (see table B.1). This situation contrasts with the complex media required for other trypanosomatids. A lack of requirement for haemin, from which, for example, the prosthetic groups of cytochromes and other enzymes are synthesised, is a further feature unique to *C. oncopelti* among the trypanosomatids. Another unusual feature of the organism is the presence of basophilic, rod-shaped particles in the cytoplasm (figure C.1). *C. oncopelti* contains one or two of these structures which are located at the posterior end of the cell. When they were discovered in 1957, they were named bipolar bodies because of their characteristic bipolar appearance in Giemsa-stained preparations. They were shown to be composed mainly of ribonucleoprotein but no suggestion was made at the time as to what they represented.

C.2 Initial Conclusions

Two distinct pathways are known to occur in nature for the biosynthesis of the amino acid lysine. α-ε-Diaminopimelic acid is a key intermediate in one of the pathways (present in bacteria, algae, plants and a few fungi) and α-amino-adipic acid in the other (present in most fungi and euglenids). Animal cells and most protozoa are unable to synthesise lysine. However, in 1962, it was reported that *C. oncopelti* synthesises lysine by the diaminopimelic acid pathway, which, as indicated above, is characteristic of bacteria. Diamino-pimelic acid decarboxylase, the terminal enzyme in this pathway, was concentrated in that fraction of the broken cell preparation of the organism which contained the bipolar body. If *C. oncopelti* was grown in a complex medium containing penicillin G (6 mg/ml), the bipolar bodies disappeared. It was suggested therefore that bipolar bodies are endosymbiotic bacteria which synthesise metabolites such as lysine for the host cell. Such a phenomenon would explain the simple nutritional requirements of *C. oncopelti*. This view was strengthened by a finding in 1963 that a DNA component, which could be distinguished, by means of its density in CsCl gradients, from those in the nucleus and kinetoplast of the organism, is associated with the bipolar bodies. Further support came from a report that the bipolar bodies in *C. oncopelti* disappeared after repeated subculture in a complex medium containing haemin. They could not, however, be cultured free from the protozoan.

C.3 Doubts

The conclusion that the bipolar bodies are endosymbiotic bacteria was questioned, however, when further evidence became available. First, it was shown that diaminopimelic acid decarboxylase was present in *C. fasciculata*, which does not contain bipolar bodies. Secondly, it became apparent that penicillin was not causing the elimination of the bipolar bodies but merely their apparent disappearance by reduction of their staining intensity. When organisms were transferred back to drug-free media, they were soon detectable again. Thirdly, it became increasingly clear that unique varieties of DNA exist in other cell organelles such as chloroplasts and mitochondria. Fourth and last, other workers failed to eliminate the bipolar bodies from *C. oncopelti* using the original technique. Thus, when a comprehensive literature review was made in 1968, it was concluded that there was no unequivocal evidence that the bipolar bodies were of bacterial origin.

C.4 The Resolution of the Problem

However, in 1971, a re-examination of the ultrastructure of the bipolar body by electron microscopy showed that it was limited by two unit membranes mutually separated by a lucent zone and that focal outpouchings into the cytoplasm occurred (figure C.2). This outer structure is similar to that of protoplasts (cell wall material removed by enzyme digestion) of Gram-negative bacteria showing everted mesosomes.

Then, in 1974, it was reported first that the antibacterial antibiotic, chloramphenicol, affected the ultrastructure of the bipolar body and subsequently that bipolar bodies could be eliminated from cultures by a single one-month period of incubation with the drug. The bipolar-body-free strain grew more slowly than the original strain, suggesting that the bipolar bodies did indeed supply nutrients to the protozoa. Also at this time, methods were developed to prepare relatively clean preparations of bipolar bodies from cells disrupted by immune lysis or sonication.

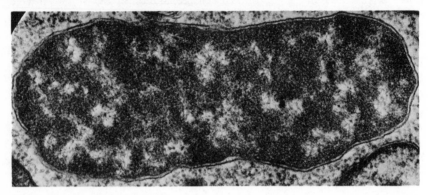

Figure C.2 Electron micrograph of the bipolar body in *Crithidia oncopelti* (× 30 000).

The bipolar-body-free strain was shown to require haemin for growth. The requirement could be replaced by protoporphyrin IX, indicating that the flagellate contains ferrochelatase, the terminal enzyme in the haem biosynthetic pathway, but not by δ-aminolevulinic acid or porphobilinogen (see figure C.3). Microassay of uroporphyrinogen synthetase revealed that the specific activity was high in the strain containing bipolar bodies, higher still in isolated bipolar bodies but negligible in the bipolar-body-free strain. It was concluded therefore that the bipolar body augments a very limited (or even non-existent) haem biosynthetic capacity of host flagellates by supplying uroporphyrinogen synthetase and perhaps other enzymes preceding ferrochelatase in the haem biosynthetic chain.

Further work showed that the DNA isolated from purified bipolar body preparations had renaturation kinetics indicative of a genome complexity similar to that in the bacterium *Escherichia coli* and far higher than that of any organelle DNAs, including kinetoplast DNA, which have so far been examined. It was also found that the bipolar bodies contain 67S ribonucleoprotein particles (compared with 87S for cytoplasmic ribosomes) and RNA components of identical molecular weights to those in *E. coli* and quite distinct from cytoplasmic rRNA (see table 6.2).

Thus, it is now generally agreed that the bipolar bodies in *C. oncopelti* do in

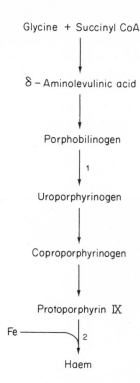

Figure C.3 Pathway for the biosynthesis of haem. Enzymes: 1, Uroporphyrinogen synthetase; 2, ferrochelatase.

fact represent endosymbiotic bacteria, their lack of sensitivity to penicillin being due to the absence of cell wall material since they are present as protoplasts. Their use in biochemical studies of parasitic trypanosomes is thus best avoided and care should be taken in the interpretation of experiments in which they have been used.

C.5 Further Reading

Chang, K-P. (1974). Ultrastructure of symbiotic bacteria in normal and antibiotic treated *Blastocrithidia culicis* and *Crithidia oncopelti. J. Protozool.,* **21,** 699–707

Chang, K-P. (1975). Reduced growth of *Blastocrithidia culicis* and *Crithidia oncopelti* freed of intracellular symbiotes by chloramphenicol. *J. Protozool.,* **22,** 271–276

Chang, K-P., Chang, C. S. and Sassa, S. (1975). Heme biosynthesis in bacterium-protozoon symbioses: enzymic defect in host hemoflagellates and complementary role of the intracellular symbiotes. *Proc. natn. Acad. Sci. U.S.A.,* **72,** 2979–2983

Newton, B. A. (1968). Biochemical peculiarities of trypanosomatid flagellates. *A. Rev. Microbiol.,* **22,** 109–130

Spencer, R. and Cross, G. A. M. (1975). Purification and properties of nucleic acids from an unusual cytoplasmic organelle in the flagellate protozoan *Crithidia oncopelti. Biochim. Biophys. Acta*, **390,** 141–154

Tuan, R. S. and Chang, K-P. (1975). Isolation of intracellular symbionts by immune lysis of flagellate protozoa and characterisation of their DNA. *J. Cell Biol.*, **65,** 309–323

Vogel, H. J. (1965). Lysine biosynthesis and evolution. In *Evolving Genes and Proteins* (ed. V. Bryson and H. J. Vogel), Academic Press, New York, pp. 25–40

Appendix D Biochemical Protozoology Literature

D.1 Literature Searches

Now that you have read this book, you might be interested in following some of the topics further. The suggestions for further reading given at the end of each chapter have been selected with this in mind since most contain a detailed list of references. Consultation of these and the literature that they cite makes it possible to build up quickly a comprehensive list of papers in any given area.

This process can of course be short-circuited, if funds are available, by the commission of a search by one of the commercial organisations which specialises in this field. Alternatively, the searching can be done personally, using either Chemical Abstracts (Chemical Abstracts Service, Columbus, Ohio) which covers most of the biochemical protozoology literature or Protozoological Abstracts (Commonwealth Agricultural Bureaux, Slough, England) Personal searches can be cross checked against advanced texts (see section D.3) and reviews that appear from time to time in proceedings of International Protozoology and Parasitology Congresses, *Annual Reviews of Biochemistry, Annual Reviews of Microbiology* (both Academic Press, New York) and some of the parasitology journals (especially *Experimental Parasitology*—Academic Press, New York).

D.2 Current Awareness

A limitation of all literature search procedures is that there is a lag of up to twelve months between the publishing of a paper and its appearance in these systems. Current awareness searches are available commercially but tend to be expensive. Most people therefore resort to either scanning the title pages of journals or consulting *Current Contents* (Life Sciences section) (Institute for Scientific Information, Philadelphia). An indication of the journals in which you are most likely to find papers in the area of biochemical protozoology is given in table D.1. Note the wide spread of biochemical, pharmacological, cytological, parasitological and microbiological journals involved. Top of the list is the *Journal of Protozoology*, the official journal published by the International Society of Protozoologists. This society is based in the U.S.A. but sections exist in many other countries (for example, United Kingdom, France,

Table D.1 Source of references for the biochemical protozoology literature*

Position	Journal	% of total references
1	*Journal of Protozoology*	9.7
2	*Biochimica Biophysica Acta*	6.6
3	*Experimental Parasitology*	5.9
4	*Nature*	5.2
5	*Annals of Tropical Medicine and Parasitology*	4.1
6=	*Comparative Biochemistry and Physiology*	3.8
6=	*Journal of Parasitology*	3.8
8	*American Journal of Tropical Medicine and Hygiene*	3.5
9	*Parasitology*	3.2
10	*Biochemical and Biophysical Research Communications*	3.0
11=	*Archives of Biochemistry and Biophysics*	2.3
11=	*Biochemical Pharmacology*	2.3
11=	*Zeitschrift für Tropenmedizin un Parasitologie*	2.3
14	*Transactions of the Royal Society of Tropical Medicine and Hygiene*	2.2
15	*Biochemistry*	2.1
16	*Proceedings of the National Academy of Sciences of the U.S.A.*	2.0
17	*Science*	1.9
18	*Journal of Biological Chemistry*	1.7
19	*European Journal of Biochemistry*	1.6
20=	*Journal of Cellular Biology*	1.5
20=	*Journal of Medicinal Chemistry*	1.5
20=	*Revista do Instituto de Medicina Tropical de Sao Paulo*	1.5
23=	*International Journal of Parasitology*	1.4
23=	*Journal of Bacteriology*	1.4
25=	*Biochemical Journal*	1.3
25=	*Molecular Pharmacology*	1.3
27=	*F.E.B.S. Letters*	1.0
27=	*International Journal of Biochemistry*	1.0
27=	*Journal of General Microbiology*	1.0
27=	*Journal of Molecular Biology*	1.0
27=	*Antimicrobial Agents and Chemotherapy*	1.0
—	All other journals (<1% each)	17.9

* Based on a survey of 954 references indexed as being of interest to biochemical protozoologists (though not necessarily directly on protozoa) during the period January 1970 to July 1976

Italy, Czechoslovakia). Membership of the society is to be encouraged since not only does it make it possible to obtain the *Journal of Protozoology* at a reduced rate but attendance at its meetings enables one to be aware of the progress being made in our understanding of biochemical protozoology long before information is published in journals.

D.3 Advanced Texts

Kidder, G. W. (ed.) (1967). *Chemical Zoology, vol. I, Protozoa.* Academic Press, New York

Lehninger, A. L. (1975). *Biochemistry*. Second edition, Worth Publishers, Inc., New York

Van den Bossche, H. (ed.) (1972). *Comparative Biochemistry of Parasites*. Academic Press, New York

Van den Bossche, H. (ed.) (1976). *Biochemistry of Parasites and Host–Parasite Relationships*. Elsevier-North Holland Biomedical Press, Amsterdam

Von Brand, T. (1973). *Biochemistry of Parasites*. Second edition, Academic Press, New York

Index

No attempt has been made to produce a comprehensive index since we regard this more as a book to read than as a work of reference. However, there is a full list of contents at the beginning of the book and the text is cross-referenced extensively. Only information which cannot be found readily from these sources has been included in this index.